Math
Magic
for Your
Kids

Also by Scott Flansburg:

Math Magic: The Human Calculator Shows How to
Master Everyday Math Problems in Seconds

Math
Magic
for Your
Kids

Hundreds
of Games and Exercises
from the Human Calculator
to Make Math Fun
and Easy

Scott Flansburg

HarperPerennial
A Division of HarperCollinsPublishers

First HarperPerennial edition published 1998.

Book design and illustration by Jennifer Harper

Library of Congress Cataloging-in-Publication Data

Flansburg, Scott.
    Math magic for your kids : hundreds of games and exercises from the human calculator to make math fun and easy / Scott Flansburg. — 1st HarperPerennial ed.
        p.   cm.
    Originally published: New York : Morrow, c1997.
    ISBN 0-06-097731-0
    1. Mathematics—Study and teaching (Elementary). I. Title.
    [QA135.5F532   1998]
    372.7'2044—dc21   97-34764

98 99 00 01 02 RRD 10 9 8 7 6 5 4 3 2 1

To my grandmothers, Ann Thornton
and the late Audrey Flansburg.
I miss you both every day.

And to my nieces and nephew,
Jennifer, Randi Marie, Jessica, and John, Jr.
Don't you dare get anything less than an A in math, or else.

Preface

Welcome to the world of the Human Calculator.

I wrote this book for children ages five to eight, but it can also be of benefit for younger children who show readiness for its number concepts and for older children who need to reinforce their basic math skills.

The primary purpose of *Math Magic for Your Kids* is to provide a positive experience of math. In fulfilling that purpose, several specific objectives will be met:

- Reinforcing the basic operations of addition, subtraction, multiplication, and division

- Introducing strategies to make working with numbers easy

- Laying a foundation for basic number sense

- Reinforcing other math concepts such as shapes and measuring

- Relating math concepts to skills of language, logic, social studies, art, etc.

- Demonstrating ways in which math is an integral part of everyday life

- Showing how much fun and how rewarding mastery of math concepts can be

- Encouraging independent thinking

- Promoting self-esteem

Math Magic for Your Kids is intended as a supplement to enrich your child's instructional experiences in the classroom. It is not a textbook, nor is it meant to serve as your child's only exposure to mathematics. Rather, it is designed to fit in with various mathematics curricula that have been approved by the National Council of Teachers of Mathematics as meeting or surpassing its standards for instructing children nationwide.

Finally, you are urged to think of this book as one more tool to assist your child in maximizing his or her potential. Together, you, your child, your child's teacher, and I can make a great team. Enjoy *Math Magic for Your Kids* and its ideas and, more than anything, use it to have fun with math while you have fun with your child's learning!

I look forward to hearing all about your successes with *Math Magic for Your Kids*. Write to me at: The Human Calculator, c/o William Morrow and Company, Inc., 1350 Avenue of the Americas, New York, N.Y. 10019.

— SCOTT FLANSBURG, the "Human Calculator"

Acknowledgments

There are a lot of people who have helped me get my books and my business, The Human Calculator, Inc., off the ground, and to whom I owe many thanks:

My sister Cindy, who got me started on the right path.

My family, who always had the same phone number.

Sue Colwell, the first person who believed in me and who also helped me get started.

Kay Dean, the second person to believe in me.

Don Davenport, who was probably the first person to realize that I might actually have something here.

Mike Levy and the crew at Positive Response, who made my program so successful.

The people at William Morrow and HarperCollins, who gave me the chance to spread my message.

Karen and Deena from Just Write in San Diego, who sat with me all those times to help me communicate my message.

Paul Bader, salesman of the millennium, and his wife, Debbie, who are always there as friends and consultants in my never-ending journey as "Math Guy."

Michael Hutchinson and his wonderful family, who created my life map.

Ray Manzella, who showed me the way.

John and Joyce Jensen, the coolest math teachers on the planet, who guided me down that mathematical path to profound enlightenment.

Jon Lovitz, who showed me the ropes.

Pat O'Brien (now if I could just dress as cool as you).

Rob and Jill Lester, two of the best people in my life, who are best friends and parents, and their three wonderful children, Kim, Stephanie, and of course my godson, Robbie, who is giving me a second childhood.

The guys of the BAIB tour, Vince Marold, Anthony Santa Maria, Mike Rowe, Bruce Hammrich, and Clive McCorkell, who shared one of the funniest years of my life.

Tony and Becky Robbins, who have made an immeasurable impact on my life, and their incredible staff, including Joseph, Brooks, Caroline, and Albert, who showed me that math and success aren't everything in life.

My golfing buds, who always let me keep score: Freeman Theriault, Brian Pavlet, Alice Cooper, Eddie Webb, and Stan Zucker.

All the math people around the world who invite me to their schools and special programs, especially Math Counts.

All my friends in radio and television who have invited me to be a part of their programs and who have provided fun, friendship, and support.

Finally, I would like to thank all of the children, parents, and teachers with whom I have been fortunate enough to share my love, enthusiasm, and passion. Every time I see someone overcome anxiety with math, it gives me the energy I need to continue my mission.

Contents

1

Introduction for Parents

This book is designed to be "user-friendly"—appealing and instructive for both children and their parents. While the instruction in this book is based on solid, proven methods, some conventional techniques, such as "carrying" in two-digit addition or subtraction, have been omitted in favor of other strategies. Also, since the focus of this book is to enrich a child's mastery of arithmetic with whole numbers from 1 through 12, neither fractions nor negative numbers are used.

To ensure that your child receives the maximum benefit from *Math Magic for Your Kids,* we recommend that you follow these guidelines:

- If your child does not read, spend twenty to thirty minutes a day (an average of three pages) reading this book with him or her. If your child does read, spend fifteen minutes a day reviewing the activities with him or her.

- Set goals with your child of completing certain pages or sections in specified time periods. One example is to complete the addition activities for the number 5 in two 15-minute work sessions. Each time your child completes a chapter, make it a special occasion—something of which you can both be proud! Consider taking your child to the park for bicycling or skating, or fixing a favorite meal for dinner that night.

- Tell your child's teacher that you are working with your child on this book, and that it is designed to supplement rather than replace the content and experiences acquired in school. If necessary, describe some of the unique ways in which arithmetic is done in this book. Keep communication lines open with teachers, and keep it positive!

Make this experience as much fun for your child as possible, continually helping him or her to observe and profit from math concepts in everyday life. For example, you could reinforce your child's grasp of addition by having him or her help you estimate the grocery bill while shopping. Or, you can work on multiplication skills by having your child help you plan a picnic menu ("If six people are coming, and if each person will want to eat two pickles, how many should we bring?"). Specific suggestions and ideas for each number are given in every chapter, and are indicated by the Close to Home icon.

Throughout this book, the following symbols are used to help you and your child quickly identify recurring elements:

 Close to Home: Everyday activity the parent can use to build a child's math sense

 Extra Credit: Additional activity to reinforce a math concept

 Fun Fact: Information that relates a math concept to other areas of learning

 Grouping: Way of relating math concepts a child already knows to those still being learned

 Hint: Principle that facilitates learning

 Journal: Writing activity that reinforces a math concept

 Magic Act: Shortcut method or activity that reinforces a math concept; often seems like a "magic trick"

 Note to Parents: Information to assist the parent(s) in extending a child's learning

 Toughie: Challenging activity to reinforce a math concept

 Trick Question: Humorous riddle that often depends on word play or logic

2
Let's Meet the Human Calculator

Introduction for Children

Hi, I'm Scott Flansburg.

Some people call me the Human Calculator. That's because I can add up numbers in my head faster than a calculator. In fact, when I was a boy, I started playing a game whenever my dad and I went to the grocery store: I'd add up everything we'd bought before the person at the cash register could. I'd add in the tax, then tell my dad the total. He'd tell the checkers, and they'd be amazed!

Actually, as you will soon find out, adding is one of the simplest things you can do. And once you know how to add, then you can also do other things like subtract, multiply, and divide. There are so many ways to make it easy, and that's what I want to share with you in this book. To help me do that, I'd like to introduce you to my buddies, Cal and Callie. They are two of my very best friends, and I hope that by the time you're done with this book, they will be among your best friends, too!

Let's Meet Cal and Callie

Here's what you can expect from us. We'll introduce you to some of our other friends, the Number Gang: the numbers 0–12. In Chapter 1 you will learn how to add them to one another, step-by-step, so that it's simple and fun. Once you've seen how we do it, you'll get the chance to do some exercises yourself. Then we'll show you a lot of tricks for making addition even easier. Then you're ready to move on to the other chapters and learn even more.

Good luck, and remember to have fun!

Let's Meet the Numbers

 Are you ready to get started?

 Good, because so are we—and do we ever have a lot of things planned for you!

Like going to camp . . . and visiting outer space . . .

 And exploring under the ocean. We'll even go back-ward — and forward — in time!

 But first, we want you to meet our best friends — the numbers from 0 to 12. We hope you like them as much as we do.

 You'll find that all together, we're quite an adventurous bunch!

Zero is a really handy guy to know. Sometimes he's kind of shy about being a "big noth-ing," but I tell him to be proud of it. In fact, Zero is lots of fun when you're adding or taking away numbers. And when it's time to multiply, that's when Zero really shows his stuff! Right now he's learning all about karate.

I have to say that **One** is one of my oldest buddies. He's really friendly and a good sport, and he gets along with all the other numbers. We like to ride our bikes together. He has a special kind of cycle called a unicycle, which has only one wheel. What a trip! If he were a bear, he could ride it in the circus!

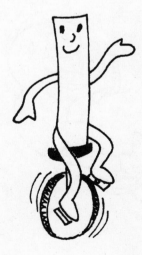

Time to meet **Two**! She moves around a lot because she likes to help all the other numbers "double" their fun. Some of the things she likes to do are dance, bicycle, and skate—and let's not forget some of her favorite water sports, like canoeing and water skiing!

You'll like **Three**. He's a nice round number who's a little bit on the shy side, but there couldn't be anybody nicer. We're all helping him to play more sports and games—it's so much fun to be active! I bet that in no time, you'll be shouting, "Three cheers for Three!"

Meet **Four**. You can usually find him in the lead, whatever he's doing. That's because he's a real "forward" thinker, always "forging" ahead! He does just about everything—golf, scuba diving, hiking, and even fencing.

Five is one of the happiest numbers we know. She's so upbeat all the time, we call her High Five. She's real outgoing and hangs around a lot with Zero and Ten. Jumping is one of her favorite activities, and she's a real fancy diver, too.

And now, meet **Six**. She's an expert at gymnastics, and some of her best buddies are Three and Nine. Everybody likes her except for one little thing: She got a guitar for her birthday, but won't take lessons. (She really needs them.) So we all run and hide when Six starts singing!

Seven is a really lucky guy; in fact, you might say that good luck just seems to follow him everywhere. He's kind of a western dude who spends a lot of time in the saddle. He's so friendly and easygoing, I know you'll like him as much as I do!

Wow, Crazy **Eight** is at it again! This guy is willing to try anything, isn't he? But here's something that's important to know: He makes sure to learn as much as he can about something before he tries it. That way he sets a good example for his other number friends. Hang in there, Eight!

I'd like you to meet **Nine**. She's a pretty deep thinker, always looking at the stars, and wants to be an astronomer when she grows up.

Calling all dudes! When you "hang ten" with **Ten**, you do it with the best! No matter what he does, this guy is just too cool. Zero is his little bro', so Ten kind of looks out for him (so does Five). Ten is superconfident, and I hope some of it rubs off on Zero.

Eleven is an all-around outdoors guy. If it calls for fun in the sun—biking, hiking, whatever—count him in! As you'll see later, he's really popular for multiplying other numbers.

I think **Twelve** is tops! He's a terrific singer and player—in fact, we call him Twelvis because Elvis Presley is his hero—and he always seems to end up in the spotlight. (Maybe he could give his little sis, Six, a few singing pointers.)

So there you have it—all the fun-loving numbers from Zero to Twelve! We can hardly wait for you to share our adventures. . . .

 And find out how easy it is to add, subtract, multiply, and divide.

 See you on the next page!

3

Putting It All Together

Addition with the Human Calculator

Adding is one of the simplest things you can do. Think about it: Suppose you're home alone and then a friend comes over. Before your friend arrived, there was just one of you; now there are two. That's adding.

0. Getting to Know...Zero!

Zero may look like just a big nothing, but you'll soon find out he's pretty important! Just remember that whenever you add any number together with 0, the answer is the same as the number you started out with. *Whenever you add zero to something, it stays the same.*

Household Pets

8 + 0 = 8

0 + 0 = 0

12 + 0 = 12

1 + 0 = 1

6 + 0 = 6

2 + 0 = 2

9 + 0 = 9

4 + 0 = 4

3 + 0 = 3

5 + 0 = 5

11 + 0 = 11

7 + 0 = 7

10 + 0 = 10

Zero Activity:

Color the Zerosaurus Rex

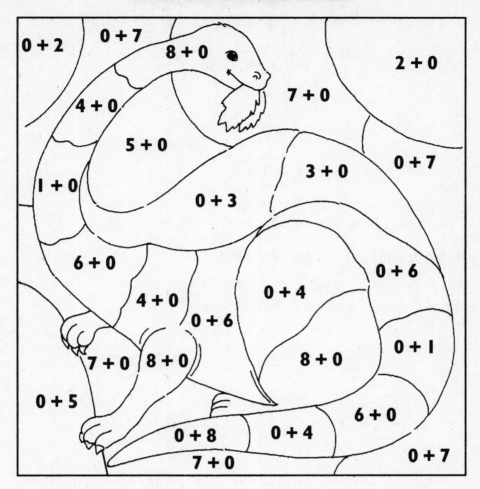

Add the numbers in each section. Then color each section according to this key: If the sum is 1, then color it **green**.

2	blue	3	red
4	green	5	blue
6	green	7	blue
8	green		

Become a Mathlete! Exercise with Cal!

Notes for Myself . . . Just for Fun!

Some people say that the number 0 reminds them of an egg. What does the number 0 remind you of?

Did you already figure out that no matter how many times you add 0 to itself, it still equals 0?

0 + 0 + 0 + 0 + 0 + 0 + 0 + 0 + 0 + 0 + 0 + 0 + 0 + 0 + 0 + 0 = 0!

Help your child understand that 0, or nothing, can be considered the same as empty space. (1) Give him/her a full jug of juice or water from the refrigerator, together with an empty jug, and ask which one has 0 in it. (2) When you come home from the store, give your child various empty storage containers and ask him or her to fill them by emptying boxes of cereal, beans, pretzels, or pasta into the containers. Now it's the packages from the store that have 0 in them! (3) Have your child identify other examples of empty versus filled spaces, such as an empty toy box versus a filled one, or the inside of an inflated balloon versus a filled water balloon. Then ask your child, "If all the empty spaces were put together, what would they amount to?"

 What I already know about zero:

What I like best about zero:

1. Fun with One

Next to 0, adding 1 is the easiest thing! If you know how to count, you can add 1 in a snap. Anything plus one is just the next number. *When you add one, count to the next number.*

Household Pets

One Activity: Count Your Toys

Write the number of toys in each group in the space provided.
Then add them together.

Example:

2 + 1 = 3

Now try it yourself:

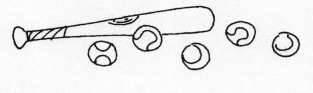

_____ + _____ = _____

_____ + _____ = _____

_____ + _____ = _____

_____ + _____ = _____

_____ + _____ = _____

Become a Mathlete! Exercise with Cal!

Schoolyard Games

10 + 1 =

11 + 1 =

2 + 1 =

6 + 1 =

7 + 1 =

8 + 1 =

5 + 1 =

9 + 1 =

1 + 1 =

0 + 1 =

4 + 1 =

3 + 1 =

12 + 1 =

Notes for Myself . . . Just for Fun

Have you ever noticed all the numbers around you? For the rest of the day, see all the numbers around you. What numbers do you see on TV? on the playground? Do any of the books you're reading have numbers in them? How about your favorite music?

Reinforce both your child's counting skills and his/her grasp of "one more" by encouraging him/her to help you "take inventory" of things around the house. (1) For younger children, you might ask them to keep track of how many beds, sinks, windows, doors, etc. there are. (2) For older children, ask them to inventory books, toys, shoes, articles of clothing, dishes, food items, etc. (3) Older children can also sort many of the above items for you by identifying relevant categories (such as their own socks, sister's shoes, Gypsy's chew toys, etc.) and putting everything in its proper place.

 What I already know about one:

What I like best about one:

2. It Takes Two to Tango

When you add with two, just skip over the next number and you're there! Adding two is like counting all the even numbers—like 2, 4, 6, 8, 10, 12, and so on—or counting all the odd numbers—like 1, 3, 5, 7, 9, 11, and so on. Piece of cake!

Two Activity:

Two by Two

Once upon a time there was an old guy named Noah. He lived where it rained a lot, so he decided to create a floating zoo. He called it Noah's Ark 'cause he thought that had a pretty good ring to it. Then he got together two of every animal. If he put on his boat 2 hippos, 2 giraffes, 2 gorillas, 2 snakes, 2 dogs, 2 cats, and 2 fleas, how many animals did he have altogether?

Become a Mathlete! Exercise with Callie!

Schoolyard Games

10 + 2 =

11 + 2 =

2 + 2 =

6 + 2 =

8 + 2 =

5 + 2 =

7 + 2 =

1 + 2 =

0 + 2 =

9 + 2 =

4 + 2 =

3 + 2 =

12 + 2 =

Notes for Myself . . . Just for Fun

Look in a mirror: What do you have two of?

Imagine you meet two space creatures from the Planet Two. All Twotians have a twin. Draw a picture of what you think these people look like, and make up a story about them. What are their favorite flowers? (Maybe tulips!)

(1) Ask your child for examples of even numbers in the house. For example, if you have a dog or cat, ask your child about its number of legs; is it an even or an odd number? How about its tail? eyes? tongue? How about your child's age, or that of other family members? (2) Reinforce your child's concept of pairs. Does anyone in the family wear glasses or contact lenses? On laundry day, talk about pairs as your child helps you match socks. If your child cannot tie shoes yet, help him or her do this as another way of conveying the concept of pairs. (3) For very young children, help them distinguish between right and left, such as shoes or sides of the body. Which wearable items must be matched to a right side or a left side (e.g., shoes)? Which are interchangeable (e.g., socks, earrings)?

What I already know about two:

What I like best about two:

3. Three's a Crowd

Become a Mathlete! Exercise with Cal!

Household Pets

$$8 + 3 = 11$$
$$0 + 3 = 3$$
$$1 + 3 = 4$$
$$12 + 3 = 15$$
$$6 + 3 = 9$$
$$9 + 3 = 12$$
$$4 + 3 = 7$$
$$2 + 3 = 5$$
$$3 + 3 = 6$$
$$5 + 3 = 8$$
$$11 + 3 = 14$$
$$7 + 3 = 10$$
$$10 + 3 = 13$$

Schoolyard Games

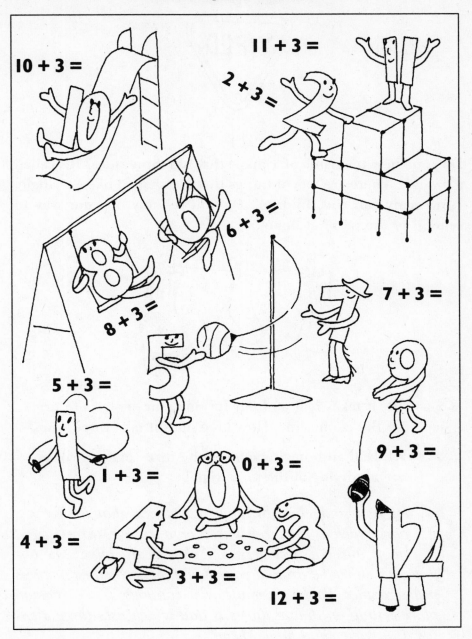

10 + 3 =

11 + 3 =

2 + 3 =

6 + 3 =

8 + 3 =

7 + 3 =

5 + 3 =

1 + 3 =

0 + 3 =

9 + 3 =

4 + 3 =

3 + 3 =

12 + 3 =

Notes for Myself . . . Just for Fun

Do you know the name of a shape that has three sides? It's called a *triangle*. There are lots of things that are shaped like a triangle. How many can you think of? Do you see any on your way to school? to the park? at home?

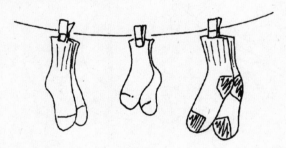

Q: It takes half an hour for one pair of socks to dry on the clothesline. How long does it take three pairs?

A: Half an hour (remember the three pairs of socks are all drying at the same time!).

(1) Ask your child to point out objects that have three parts, such as the three wheels on a tricycle, three-legged stools, three-leaf clovers, or "cloverleaf" buns. The objects can either be around your house, or located as pictures in magazines, books, or mail-order catalogs. (2) A common place-setting on dining tables is that which uses three utensils. Can your child name them?

 What I already know about three:

What I like about three:

If your little sister Katie has only 1 tooth, then gets 1 more, how many teeth would she have?

What if she got 2 more after that? Then how many would she have?

This is the same as saying:

```
  1
  1
+ 2
  4
```

But you don't have to add all three numbers at once. You can start with any two of the numbers you want—the easiest ones for you—then add in the other number. For example, you could start with

```
  1
 +1
  2
```

Then do:

```
  2
 +2
  4
```

Or you could do it this way:

```
  1
 +2
  3
```

Then do:

```
  3
 +1
  4
```

Three Activity:  "Grouping"

Now you try it. Suppose you catch 1 ladybug. Then your best friend brings you 2 more that he has caught. Then his sister brings you 3 that she has caught. When you put them all in a jar, how many ladybugs do you have all together?

This is the same as saying:

$$
\begin{array}{r}
1 \\
2 \\
+\ 3 \\
\hline
\end{array}
$$

How would you figure this out? Which two numbers are easiest for you to add together first?

Or how about this one? Your mom gives you 3 pieces of bubble gum. Then your dad gives you 2 more. You find 1 more piece under your bed. Your friend has 2 more, so you stick them all in your pocket and head for school. How many pieces of gum do you have to share with your friends?

This is the same as saying:

$$
\begin{array}{r}
3 \\
2 \\
1 \\
+\ 2 \\
\hline
\end{array}
$$

Which two numbers would you start with? Remember, there is no one right answer. Just start with the two numbers that are easiest for you to add, then move on to the next.

What if you completely forgot about the gum, and put your pants in the dirty clothes basket when you got home? After your pants have been washed, and you check your pockets, how many pieces of gum would you have?

One! Just one big gooey mess!

4. Four!

Household Pets

Four Activity:

Hit the Bases

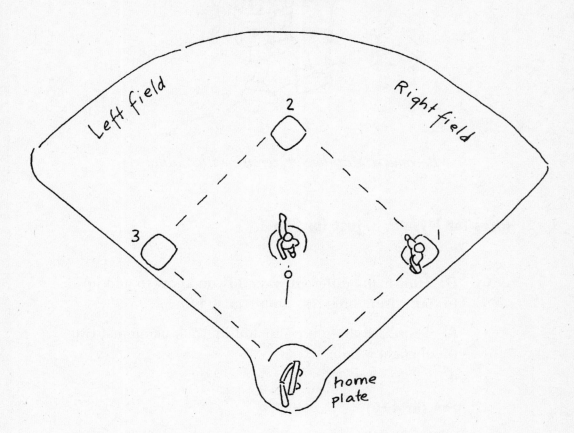

Cal is at bat. He already has one teammate on first base. If Cal hits a single, what base does his teammate advance to? If Cal hits a double, what base would his teammate go to? What if Cal hits a home run?

What if there were two teammates already on base when Cal came up to bat? Can you think of other ways his team could score?

Become a Mathlete! Exercise with Callie!

Notes for Myself . . . Just for Fun

Q: How many different ways do you know to add up to four, using only two numbers at a time?

A: There are three ways to add up to 4, using just two numbers at a time.

4

0+4 (or 4+0)

1+3 (or 3+1)

2+2

Note: Of course, there are many other ways to add up to 4 using more than two numbers at a time, for example:

1+1+1+1

2+1+1

0+1+0+0+1+0+1+0+0+0+0+1+0+0+0

Now, how many ways do you know to add up to 5, using only two numbers at a time? Write the answers below.

5

    +

    +

    +

Answer: ___ways

How about 6 and even higher numbers? Write the answers below each number.

6   **7**

Answer: ___ways   Answer: ___ways

8   **9**

Answer: ___ways   Answer: ___ways

Now look at the answer for each number. Do you notice a pattern? Without doing this activity for the number 10, what do you think the answer would be?

Help your child see that the answers for 4 and 5 are 3, for 6 and 7 are 4, and for 8 and 9 are 5.

(1) Go on a scavenger hunt with your child for different sizes of squares and rectangles. In the office or den, look for business-size and 8½-by-11-inch envelopes, postage stamps, desk-pad calendars, and picture frames. (2) Show

your child a flat, square shape and ask what he/she would do to "turn it" into a diamond (i.e., get your child to rotate it). (3) Take your child to a baseball game, whether Little League, softball, T-ball, or professional. How many bases must a player run in order to score? How many sides does a baseball diamond have?

 What I already know about four:

What I like about four:

5. Five Alive!

Household Pets

Five Activity:

Make Sense with Cents

Get some coins from your piggy bank, or ask your parents for some. Now look at the coins. How many different kinds do you have? How many cents is a penny worth? a nickel? a dime? a quarter? Do you have any fifty-cent pieces? Any silver dollars or Susan B. Anthony dollars?

Add all your coins together. (Ask someone for help if you need it.) How much are they worth? Now set aside all the pennies, and add up the coins that are left. How much are these worth?

Did you notice that when you added up all the coins that weren't pennies, your answer was a number that ended either in 0 or 5? Why do you think that is?

 Point out to your child that except for pennies, all the other coins have number values that end in 0 or 5— nickels (5), dimes (10), quarters (25), etc. Therefore, all their sums will also end in 0 or 5.

Notes for Myself . . . Just for Fun

Adding by 5s is one of the easiest things to do. You know that every other number ends in a 0, and the numbers in between end in 5. Just look: 0, 5, 10, 15, 20, 25, 30, 35, and so on.

Q: Cal has two coins in his hand. Together they add up to 6 cents. One of the coins is not a nickel. What are the coins?

A: A penny and a nickel. One of the coins is not a nickel — but the other one is!

As you shop for groceries, have your child find the price tag on items that go into your shopping cart and read the prices aloud. Pick two items for comparison and ask which one costs more. At the produce counter, if you weigh vegetables, fruits, or beans in a scale, share with your child what you are doing. As your child becomes more experienced and interested in helping you shop, he/she can help you compare prices per unit among brand names.

What I already know about five:

What I like about five:

6. Tricks with Six

Become a Mathlete! Exercise with Cal!

Schoolyard Games

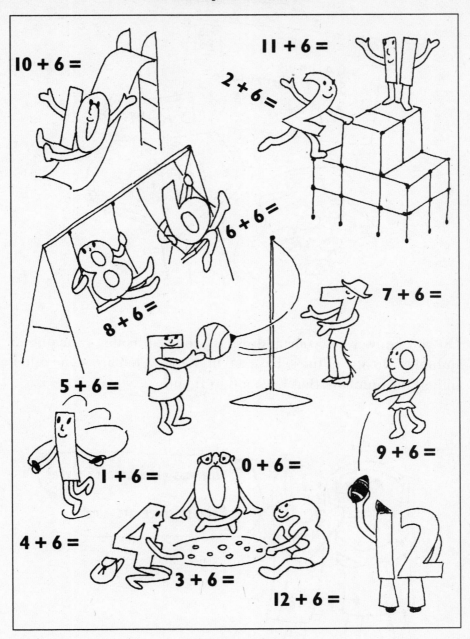

10 + 6 =

11 + 6 =

2 + 6 =

6 + 6 =

8 + 6 =

7 + 6 =

5 + 6 =

1 + 6 =

0 + 6 =

9 + 6 =

4 + 6 =

3 + 6 =

12 + 6 =

Notes for Myself . . . Just for Fun

Do you know what your address is? your zip code? your phone number? Do any of these have a 6 in them? What are some other important numbers that have a 6 in them?

Q: Why was six scared of seven?

A: Because seven ate nine!

(1) It is very important for children to know their address and phone number as soon as they are able—age four is not too young. See if you can make a rhyme, game, or song out of it to help your child remember, and reinforce it frequently. (2) Also make sure your child understands how to call for help in an emergency, and what an emergency is. Does your child know how to dial 911? (3) Stop signs are relatively easy for prereaders to grasp because they are identifiable by their color and shape. Point out to your child that a stop sign is a hexagon, a six-sided figure. Does your child understand traffic signals and know how to safely cross a street?

What I already know about six:

What I like about six:

Shortcuts for Addition: More "Grouping"

Remember how to add more than two numbers at a time? Just add the two numbers that are easiest for you; then add the other numbers, one at a time.

It's as if you're picking up your best friend first on your way to the roller rink. If you were going skating with Two, Three, Four, and Six, whom would you stop for first?

In other words, if you had to add 2, 3, 4, and 6, where would you begin? Here's one way:

$$\begin{array}{r} 2 \\ +\ 4 \\ \hline 6 \end{array}$$

Then do:

$$\begin{array}{r} 3 \\ +\ 6 \\ \hline 9 \end{array}$$

Then do:

$$\begin{array}{r} 6 \\ +\ 9 \\ \hline 15 \end{array}$$ is the answer!

Can you think of another way?

Now try this one: You have 6 stickers on your school notebook. If you go to a birthday party and get 3, then go home and get 1 from your brother, then find 4 more stashed in your desk drawer, how many do you have to stick on your binder now?

$$\begin{array}{r} 6 \\ 3 \\ 1 \\ +\ 4 \\ \hline \end{array}$$

7. Lucky Seven

Become a Mathlete! Exercise with Callie

Household Pets

8 + 7 = 15

0 + 7 = 7

1 + 7 = 8

12 + 7 = 19

6 + 7 = 13

2 + 7 = 9

9 + 7 = 16

4 + 7 = 11

3 + 7 = 10

5 + 7 = 12

11 + 7 = 18

7 + 7 = 14

10 + 7 = 17

Seven Activity:

A Few of My Favorite Things

Draw 7 of your favorite animals. Then draw 5 of your favorite toys. How many favorite things have you drawn?

Now find 2 pictures of your favorite thing to do and 7 pictures of your favorite dolls or action figures. How many favorite things have you found?

Make a list of 7 of your favorite books or stories. Now list 3 of your favorite TV shows or movies. How many of your favorites have you listed?

Schoolyard Games

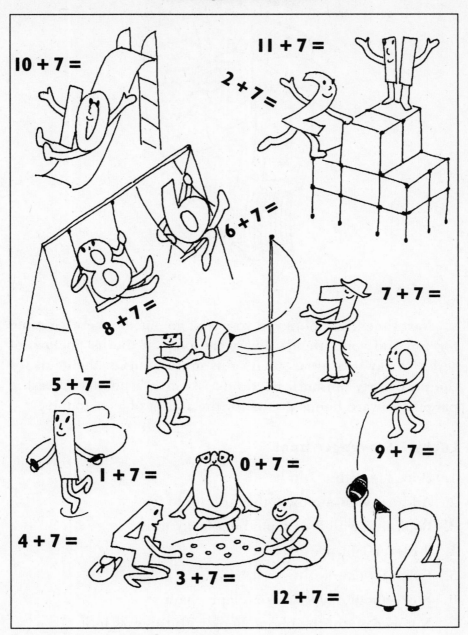

10 + 7 =

11 + 7 =

2 + 7 =

6 + 7 =

8 + 7 =

7 + 7 =

5 + 7 =

1 + 7 =

0 + 7 =

9 + 7 =

4 + 7 =

3 + 7 =

12 + 7 =

Notes for Myself . . . Just for Fun

Next time it's raining and you can't go outside, here's a game for you and your friends. Everyone looks at the list (below) of seven "lucky treasures," then splits up to hunt for the items for the next thirty minutes. Each one is worth 7 points. (Or ask a parent to award bonus points for the silliest objects found.)

Lucky 7 Scavenger Hunt

1. Price tag with a 7 in it
2. Address with numbers that add up to 7
3. Book title with 7 or something lucky in it
4. 7-Eleven Slurpee
5. Calendar date with 7 in it
6. Story about or picture of a leprechaun
7. Number out of the phone book with three 7s in it

(1) On a map of the United States, show your child what city and state you live in. On a globe, show your country and continent. Does your child know how many continents there are? Review the seven continents with your child, pointing them out on the globe. Also show where the equator is and how it divides the globe into Northern and Southern hemispheres. (2) For older children, a quick tour of a street atlas might be helpful. Use one to show your child where your street is. Show how the street is located in a particular quadrant of the map (for example, A6) and how you trace vertically and horizontally to arrive at that quadrant.

What I already know about seven:

What I like about seven:

8. Crazy Eight

Household Pets

8 + 8 = 16

0 + 8 = 8

1 + 8 = 9

12 + 8 = 20

6 + 8 = 14

9 + 8 = 17

2 + 8 = 10

4 + 8 = 12

3 + 8 = 11

5 + 8 = 13

11 + 8 = 19

7 + 8 = 15

10 + 8 = 18

Eight Activity:

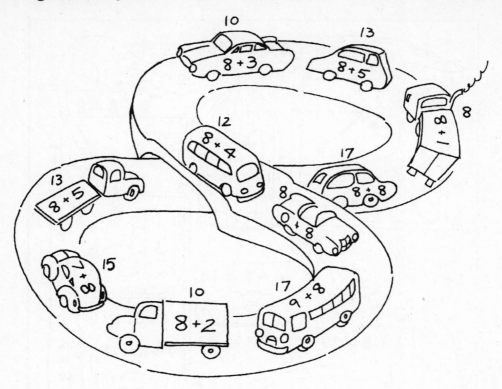

Don't Be Late . . . Race with Eight!

Another name for the answer to an addition exercise is "sum total." Add the numbers on the side of each of the race cars to see if the sum total equals the number above each car. Circle the cars whose totals do match; draw an "X" on the cars whose totals do not match.

Schoolyard Games

10 + 8 =

11 + 8 =

2 + 8 =

6 + 8 =

8 + 8 =

7 + 8 =

5 + 8 =

1 + 8 =

0 + 8 =

9 + 8 =

4 + 8 =

3 + 8 =

12 + 8 =

Become a Mathlete! Exercise with Cal!

Notes for Myself . . . Just for Fun

Do you know anyone who is eight years old? anyone who is eighteen? anyone who is eighty? Do any of your family members have an 8 in their age?

 Ask your child how old he or she will be in eight years. At that age, what grade will the child be in, and what kinds of activities is he or she likely to be doing? What year will it be? (Help them with this question if necessary, discussing the years after 2000 and its status as the twenty-first century.) You could also talk about cycles (how many summers will have passed? how many leap years?) and regularly occurring events (how many summer Olympics? how many presidential elections?).

What I already know about eight:

What I like about eight:

9. Nine on the Line

Nine is one of the easiest numbers to add with. Just think of it as 10, then take away 1. For example, if you want to solve 2 + 9 quickly, think of it as 2 + 10, which is 12. Then take away 1, and you have the answer: 11.

Become a Mathlete! Exercise with Cal!

Household Pets

8 + 9 = 17

0 + 9 = 9

12 + 9 = 21

1 + 9 = 10

6 + 9 = 15

2 + 9 = 11

9 + 9 = 18

4 + 9 = 13

3 + 9 = 12

5 + 9 = 14

11 + 9 = 20

7 + 9 = 16

10 + 9 = 19

Astronaut Cal is exploring the moon and needs your help making a safe journey. Some craters are safe to cross, but others are too deep. Help him figure out which craters he can safely cross by adding the numbers on each one. If the answer shown is wrong, cross it out—that's a dangerous crater! Circle the craters that show correct answers so Cal will know which way to go.

Schoolyard Games

10 + 9 =

11 + 9 =

2 + 9 =

6 + 9 =

8 + 9 =

7 + 9 =

5 + 9 =

9 + 9 =

1 + 9 =

0 + 9 =

4 + 9 =

3 + 9 =

12 + 9 =

Notes for Myself . . . Just for Fun

Think of all the different things you can do in one minute, and list them. For example, Cal can hop on one leg 120 times and count to 500 in one minute. How about you? How many toys can you pick up off the floor in one minute? How many words can you read in one minute? How many times can you bounce a basketball in one minute?

How many things can you do in nine minutes? What are some things that you can't do in one minute that you can do in nine minutes?

"Nine Squares." *Can you fill in the empty squares below? Write either a 2 or a 3 in each square so that all the rows, columns, and diagonals add up to 9.*

 Note to Parents: Explain to your child what a row, column, and diagonal are if necessary.

	4	
4		
		4

 Can your child name nine people in the immediate and/ or extended family? (Think of a friend's family, too, if you need to.) How many grandmothers are there? grandfathers? aunts, uncles, cousins? Draw a family tree with your child to illustrate the relationships of each person to another.

What I already know about nine:

What I like about nine:

Shortcuts for Addition: More "Grouping"

By now you're probably a whiz at grouping! Remember, if you're adding more than two numbers, just start with the ones you like best and go from there.

This is Sydney's family at Thanksgiving. She's trying to figure out how many people there are altogether, but they keep moving around from room to room. What are some different ways that Sydney can count how many people there are?

First she could count all her cousins (7), then add her uncles and aunts (5), then add her grandparents (3), then add her parents (2), and then add in her little sister (1). Oh, let's not forget the wacky neighbors from next door (4) and her best friend Lucy and her family (5).

Here's what Sydney needs to add:

7
5
3
2
1
4
+5
———
= _ _

Which two numbers are easiest for you to add first? Do whatever works for you!

 A trick Cal uses is to look for groups of 10 and add them all together, then add any numbers that are left over.

10. Hang Ten!

When you're adding 10 to any number from 1 to 9, just put a 1 in front of that other number. For example, 10 + 9 is 19. Way cool, dude!

Household Pets

8 + 10 = 18

0 + 10 = 10

12 + 10 = 22

1 + 10 = 11

6 + 10 = 16

2 + 10 = 12

9 + 10 = 19

4 + 10 = 14

3 + 10 = 13

5 + 10 = 15

11 + 10 = 21

7 + 10 = 17

10 + 10 = 20

Ten Activity:

Wax On, Wax Off

Any surfer knows that you have to take care of your surfboard, and Ten wants to buy some wax to polish it. Do the exercises below, then circle the one that matches the price of the wax. (Remember that a dime equals 10 cents, a nickel equals 5 cents, and a penny equals 1 cent.)

Become a Mathlete! Exercise with Cal!

Schoolyard Games

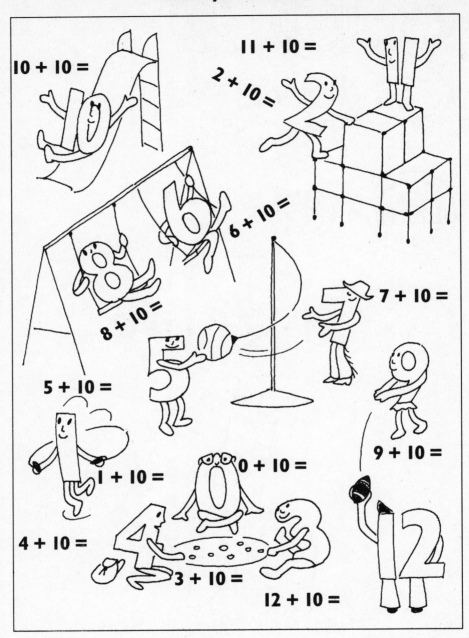

10 + 10 =

11 + 10 =

2 + 10 =

6 + 10 =

8 + 10 =

7 + 10 =

5 + 10 =

9 + 10 =

1 + 10 =

0 + 10 =

4 + 10 =

3 + 10 =

12 + 10 =

Notes for Myself . . . Just for Fun

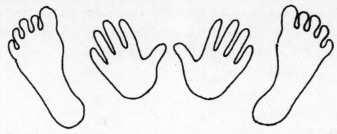

How many fingers and toes do you have? What else do you have 10 of? 10 lizards? 10 friends? 10 cookies?

Q: Kelsey and Jenna are playing tennis. Each plays 10 and wins 10 games. (Remember, in tennis there is no such thing as a tied game.) How could this happen?

A: They are playing other people, not each other.

Play "Toy Store" with your child. (1) First he or she sets up some toys around the room, then indicates their prices, such as 3¢, 5¢, 9¢, 12¢, 16¢, etc. Use pennies and dimes (or counters such as pinto beans for pennies and kidney beans for dimes) to purchase items from the store. (2) Then have the "10¢ markup" where your child marks up selected items by adding 10¢ and adjusting the price tags accordingly (3¢ becomes 13¢, etc.). (For older children, you could also invent explanations to justify the markup, such as, "The wheel factory was hit by a hurricane, and now they're behind schedule on their toy truck wheels.") Purchase the items at their new prices so that your child grasps how many coins or counters correspond to these prices and also that you can't buy as much. (3) Reverse the process and have a "10¢ off" sale. This will come in handy later for subtraction.

What I already know about ten:

What I like about ten:

Zeroes . . . Now You See 'Em, Now You Don't!

What's an easy way to add 10 + 20? Just take off the zeros. Then you have 1 + 2. What's the answer? Right! It's 3. Now put a zero at the end of the 3. You get 30, and that's the answer!

Now you try it:

$$\begin{array}{r} 30 \\ +20 \\ \hline \end{array}$$ (Hint: What's 3 + 2?)

$$\begin{array}{r} 40 \\ +50 \\ \hline \end{array}$$ (Hint: What's 4 + 5?)

$$\begin{array}{r} 40 \\ +60 \\ \hline \end{array}$$ (Hint: What's 4 + 6?)

Remember, it's okay to take out the 0, but don't forget to put it back when you're done!

11. Revvin' with Eleven

Adding with 11 is almost as easy as adding with 10. For example, if you need to add 6 and 11, just add 6 + 10, then add 1 more! What's your answer? 16 + 1 = 17!

Household Pets

8 + 11 = 19

0 + 11 = 11

12 + 11 = 23

1 + 11 = 12

6 + 11 = 17

4 + 11 = 15

2 + 11 = 13

9 + 11 = 20

3 + 11 = 14

5 + 11 = 16

11 + 11 = 22

7 + 11 = 18

10 + 11 = 21

Eleven Activity:

Martian Message

Help! Cal has stumbled on a secret message left by Martians who stopped at Earth on their way back home. To translate their language, do each exercise, then place the letter next to it on the blank matching the answer. When all the blanks are filled in, the secret message will magically appear.

3+11=A	11+2=S
11+0=F	5+1=W
6+11=H	7+11=L
4+11=U	1+11=M
3+2=T	11+9=I
12+11=O	4+0=D
5+3=R	

$$\overline{12}\ \overline{14}\ \overline{5}\ \overline{17}\quad \overline{20}\ \overline{13}\quad \overline{23}\ \overline{15}\ \overline{5}\quad \overline{23}\ \overline{11}$$

$$\overline{5}\ \overline{17}\ \overline{20}\ \overline{13}\quad \overline{6}\ \overline{23}\ \overline{8}\ \overline{18}\ \overline{4}!$$

Schoolyard Games

10 + 11 =

11 + 11 =

2 + 11 =

6 + 11 =

8 + 11 =

5 + 11 =

7 + 11 =

1 + 11 =

0 + 11 =

9 + 11 =

4 + 11 =

3 + 11 =

12 + 11 =

Notes for Myself . . . Just for Fun

How many names can you think of that rhyme with numbers? Sue, Lou, and Hugh all rhyme with 2. What names can you think of that rhyme with 11? What are some other words that rhyme with 11? What other number rhymes with 11?

 What is your child's favorite recess/playground activity? Jumping rope, swinging, climbing, running, foursquare, tetherball, or other games? Which of these can be played alone, and which require more than one person? Which can be done either way? Ask which numbers apply to the particular activity. For example, how many squares in hopscotch? How high can your child count? What numbers are used in jump-rope rhymes? Are some marbles, trading cards, or milk bottle caps worth more than others? Share with your child what your favorite playground activity used to be.

What I already know about eleven:

What I like about eleven:

Become a Mathlete! Exercise with Cal!

12. Twistin' with Twelve

Household Pets

8 + 12 = 20

0 + 12 = 12

1 + 12 = 13

12 + 12 = 24

6 + 12 = 18

4 + 12 = 16

2 + 12 = 14

9 + 12 = 21

3 + 12 = 15

5 + 12 = 17

11 + 12 = 23

7 + 12 = 19

10 + 12 = 22

Schoolyard Games

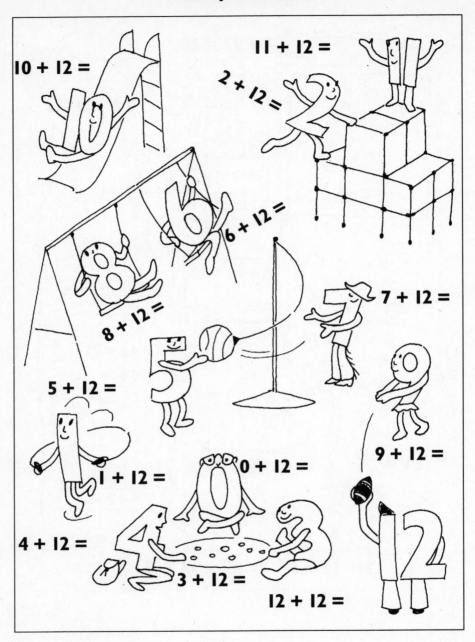

10 + 12 =

11 + 12 =

2 + 12 =

6 + 12 =

8 + 12 =

7 + 12 =

5 + 12 =

9 + 12 =

1 + 12 =

0 + 12 =

4 + 12 =

3 + 12 =

12 + 12 =

Notes for Myself . . . Just for Fun

Lots of things come in dozens.
How many can you think of?

Do you know where the idea of a "baker's dozen"—thirteen of something—came from? Long ago in England, the king ordered his royal baker to deliver twelve doughnuts to his court every day. What's more, if any doughnuts were missing, the baker would be beheaded! The baker liked his head right where it was, so he quickly started including an extra doughnut with each order of twelve, just in case.

(1) Review the twelve months of the year with your child. Does he or she know them all yet? Explain that the year is broken up into four seasons that roughly correspond to three months each; such as spring (March, April, May), etc. Which month or season is your child's favorite? (2) You can also use the twelve numbers on a clock face to introduce your child to telling time, or for reinforcement. Show your child the difference between an analog clock and a digital clock.

 What I already know about twelve:

What I like about twelve:

Left-to-Right Addition

Now you know how to add 0 through 12! You're ready to add together numbers bigger than 10.

Remember when you were adding 11 + 12? Did you start with 12, then count up to 23? Here's a shortcut, so you don't have to count anymore.

The first thing you should do is find something called the 10s column: This is the second column from the right.

$$\begin{array}{r} 11 \\ +12 \\ \hline \end{array}$$

Next, find the 1s column. This is the column on the right.

$$\begin{array}{r} 11 \\ +12 \\ \hline \end{array}$$

Got them straight? Good! Now, add the numbers in the 10s column: 1 + 1. Remember to stay only in the 10s column.

$$\begin{array}{r} 11 \\ +12 \\ \hline 2 \end{array}$$

Next, add the numbers in the 1s column: 1 + 2. Remember to stay only in the 1s column.

$$\begin{array}{r} 11 \\ +12 \\ \hline 23 \end{array}$$

Easy as pie! Now let's try two bigger numbers.

```
  24
 +31
```

"Keep those columns straight!"

First, add the numbers in the 10s column: 2 + 3. Stay only in the 10s column.

```
  24
 +31
  5_
```

Next, add the numbers in the 1s column: 4 + 1. Stay only in the 1s column.

```
  24
 +31
  55
```

Remember to stay in one column at a time.

Now you're ready to do one by yourself. Remember, all you have to do is find the 10s column and add the numbers in it, then find the 1s column and add the numbers in it.

    35
    +52

 What's in a Name?

Use the chart on page 87 to figure out how much your name is worth. Just add the values of each of the letters of your name together. For example, Cal would be:

C = **$ 3** (third letter in the alphabet)
A = **$ 1** (first letter in the alphabet)
L = **$12** (twelfth letter in the alphabet)
    $16

Now try it with your own name. Then try it with your friends' names. Remember to use what you have learned—grouping, zeros, and left-to-right addition—to make it easy!

A = $1	N = $14
B = $2	O = $15
C = $3	P = $16
D = $4	Q = $17
E = $5	R = $18
F = $6	S = $19
G = $7	T = $20
H = $8	U = $21
I = $9	V = $22
J = $10	W = $23
K = $11	X = $24
L = $12	Y = $25
M = $13	Z = $26

You can play this game with all kinds of words, such as things you see every day. (How much is your "chair" worth? What does your "bed" cost?) To have even more fun with your friends, ask one another for words that are worth a certain amount ("You can have an 'egg' for $19"), or find different words that "cost" the same.

Addition Review

Wow! Wasn't it fun putting it all together with addition? You're really a Mathlete! Now that you know how to add, you can look at two numbers—like 2 and 3—and say how many they are altogether—5. Whenever you add, your answer is called a *sum*. The sum of 2 and 3 is 5.

Lots of numbers are especially easy to add with. Whenever you add 0 to a number, for example, the number stays the same (remember, you're adding "nothing" to that number!). Adding 1 to any number is just like counting, because the answer will always be the next number (for example, $5 + 1 = 6$; and $6 + 1 = 7$). When you add 2 to any number, you skip the next number and go straight to the one after it. For example, $4 + 2 = 6$ (you skipped the next number after 4, which is 5, to get to the sum, which is 6). Adding by 2s lets you play with all *even* numbers—2, 4, 6, 8, 10, 12—or all *odd* numbers—1, 3, 5, 7, 9, 11.

Another easy number to add with is 10. When you add 10 to any number from 0 to 9, the sum will be the same except with a 1 written in front of it (for example, $4 + 10 = 14$; and $8 + 10 = 18$). And once you know how to add with 10, it's fun to add with 9, too, because 9 is the same as 10 except with 1 taken away. Let's say you want to add 9 to 3. First think of it as $10 + 3$, which gives you 13. Then take 1 away (because 9 is 1 less than 10), and you get 12—and that's the sum of $9 + 3$.

For the same reason, it's a snap to add 11 to any number. Let's say you want to add 11 to 2. First think of it as $10 + 2$, which gives you 12. Then *add* 1 (because 11 is 1 more than 10), and you get 13—and that's the sum of $11 + 2$.

You also learned about a handy shortcut called *grouping*. This lets you add more than two numbers at a time. First pick any two of the numbers you want and add them together; then add that sum to the next number, and keep going until all the numbers have been added. Let's say that you're adding 1, 2, and 3. You could start by adding 1 and 2 together first, then adding their sum to 3. Or you could start by adding 1 and 3 together, then adding their sum to 2. Or you could start by adding 2 and 3 together, then adding their sum to 1. Just choose the easiest way!

Adding numbers that end in 0, like 10 and 20, is also a cinch. First, cross out or take the 0s off (just for a little while!) and pretend that the 10 and 20 are 1 and 2. Add 1 and 2 to get the sum of 3. Now put a 0 at the end of 3 to get 30—and that's the sum of 10 and 20. Pretty tricky!

Now that you're a whiz at adding the numbers 0 through 12, you can add together numbers that are bigger than 10. Remember, here's how to do *left-to-right addition:*

$$
\begin{array}{r}
14 \\
+25 \\
\hline
\end{array}
$$

First, add 1 + 2.

$$
\begin{array}{r}
14 \\
+25 \\
\hline
3\_ \\
\end{array}
$$

Next, add 4 + 5.

$$
\begin{array}{r}
14 \\
+\ 25 \\
\hline
39 \\
\end{array}
$$

The sum is 39. It couldn't be easier!

Up and Add 'Em

How well do you know your addition? Fill in each of the squares in the table on page 90 with the correct sum, adding each number in the left column to each number in the top row, until all the squares have been filled in. Have someone time you to see how fast you did it. Then check your answers against the Addition Table on page 92. Good luck!

Addition Table

+	0	1	2	3	4	5	6	7	8	9	10	11	12
0	0												
1		2											
2		3	4										
3													
4													
5													
6													
7													
8													
9													
10													
11													
12													

Following the examples, do each of the pairs of addition exercises on page 91 to find out which is greater. If the two are the same, write an equals sign between them. If the sum on the left is bigger, write the *greater than* sign (>) between them. If the sum on the left is smaller, write the *less than* sign (<) between them.

Examples:

$$1 + 2 > 0 + 1 \quad \text{(3 is greater than 1)}$$
$$3 + 11 = 6 + 8 \quad \text{(14 equals 14)}$$
$$4 + 5 < 8 + 2 \quad \text{(9 is less than 10)}$$

1. 7 + 1   6 + 3
2. 12 + 4   7 + 8
3. 10 + wheels on a bicycle   6 + 6
4. 4 + 9   6 + 7
5. 2 + 8   7 + 4
6. 1 + 5   3 + eyes on your face
7. 0 + 11   9 + 3
8. 10 + 9   6 + 12
9. 2 + 1   1 + 2
10. 9 + 7   11 + 3
11. legs on a hippo + 4   2 + 5
12. 5 + 8   7 + 7
13. 8 + 10   6 + 11
14. 4 + 2   5 + 1
15. 2 + doughnuts in a dozen   4 + 11
16. 10 + 6   9 + 8
17. 0 + 3   1 + 1
18. 6 + 4   8 + 1
19. 9 + 11   13 + 7
20. 5 + 9   months in a year + 1

Addition Table Activities

1. Look at the Addition Table on page 92. Which sum appears most often?

2. Which sum appears least often?

3. How often does your age appear as a sum?

Addition Table

+	0	1	2	3	4	5	6	7	8	9	10	11	12
0	0	1	2	3	4	5	6	7	8	9	10	11	12
1	1	2	3	4	5	6	7	8	9	10	11	12	13
2	2	3	4	5	6	7	8	9	10	11	12	13	14
3	3	4	5	6	7	8	9	10	11	12	13	14	15
4	4	5	6	7	8	9	10	11	12	13	14	15	16
5	5	6	7	8	9	10	11	12	13	14	15	16	17
6	6	7	8	9	10	11	12	13	14	15	16	17	18
7	7	8	9	10	11	12	13	14	15	16	17	18	19
8	8	9	10	11	12	13	14	15	16	17	18	19	20
9	9	10	11	12	13	14	15	16	17	18	19	20	21
10	10	11	12	13	14	15	16	17	18	19	20	21	22
11	11	12	13	14	15	16	17	18	19	20	21	22	23
12	12	13	14	15	16	17	18	19	20	21	22	23	24

4. Use a yellow or other light-colored marker to shade in all the squares with the number 12 in them. Do you see a diagonal? If you wanted to make an X using this as one of the two diagonals, what other squares would have to be shaded? What do those squares have in common?

5. Try adding all the numbers in the 1s column above, then add all the numbers in the 2s column. How much bigger is the sum of the 2s column? What do you guess will be the sum of the 3s column? Check your guess.

4

Take It Away, Cal!

Subtraction with the Human Calculator

We sure hope you enjoyed "putting it all together" with addition. Now you have a chance to do something just as easy: subtraction.

Throughout this chapter, you'll see us giving things away to our friends. This is to help you make a picture of subtraction in your mind. Subtraction can mean giving away, taking away, putting away . . . whatever away, just so long as it's gone when you're through and you can count what remains.

You can think of subtraction as the opposite of addition. Remember, in addition you find out how many of something you have altogether. In subtraction, you see how many you have left after some are taken away. That's why we say, "Five take away 4 is 1."

Another way of figuring it out would be to ask, "What plus 4 is 5?" or "What number, added to 4, equals 5?" For some people, it's just easier to think of subtraction as negative or reverse addition.

The real question is, what's easier for you?

Good luck with subtraction, and remember to have fun!

0. Zip for Zero

Remember how easy it was to add 0 to any number? Well, subtracting 0 from any number is just as easy. The answer is always the same as what you started out with! That's because when you subtract 0, it means you're taking zero—zip, nothing, *nada*, nil—away.

Zoo Animals

Zero Activity:

Closet Cleanup

What a mess! Write each answer either with a number or by drawing how many things are left.

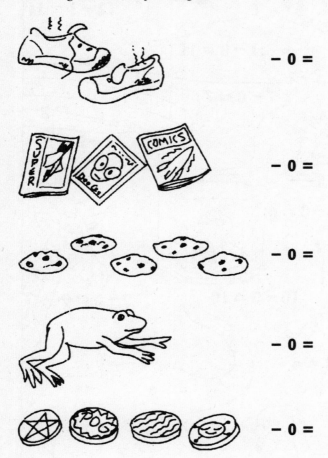

– 0 =

– 0 =

– 0 =

– 0 =

– 0 =

 Now you're an expert at taking away zero! Do you know these answers?

4,792 – 0 =

xyz **– 0 =**

Become a Mathlete! Exercise with Cal!

Notes for Myself . . . Just for Fun!

When you subtract, you take some things away, then see what is left. There are lots of different ways to subtract. For example, if you draw 2 apples on a piece of paper, then erase 1, that's one way of taking it away. Or if you draw a big X over one of the apples, that would be another way of subtracting it. How many other ways can you subtract pictures of apples?

Now, if you had some real apples in front of you, how could you make them disappear? Could you move them? hide them? eat them?

Ask your family members if they have other ideas for removing or subtracting things.

Q: Why does Cal like Zero?

A: "There's nothin' like it!"

(1) Show your child different ways in which objects or quantities can be changed or rearranged without anything being taken away. For example, if refrigerator magnets are shifted into a different pattern but are all still there, that's like taking away 0. (On the other hand, if one or more magnets were removed from the refrigerator, you'd

be subtracting something.) (2) On a larger scale, talk with your child about different ways of disposing of aluminum, glass, and paper products. Which is more like subtracting 0—mixing them with the rest of the trash, or recycling them so they can be processed and used again? If your family hasn't already done so, now would be a good time to start recycling your waste products.

 My favorite number to take 0 away from is:

Here's a trick to help me remember how to subtract 0:

1. Please Take One

Subtracting 1 from any number is a breeze! Your answer is always the number that comes just before the one you started out with. For example, the number just before 3 is 2, and guess what 3 take away 1 is? That's right, 2. (Or, to think of it another way, what number would you add to 1 to make 3? The answer is 2.)

If someone asks you, "How old were you last year?" you'd figure it out by subtracting 1. Say you're 8 years old. You take 1 away from 8 to get your answer: You were 7 a year ago.

Zoo Animals

Team Sports

9 − 1 =

1 − 1 =

11 − 1 =

2 − 1 =

8 − 1 =

4 − 1 =

6 − 1 =

10 − 1 =

5 − 1 =

12 − 1 =

7 − 1 =

13 − 1 =

3 − 1 =

Become a Mathlete! Exercise with Cal!

Notes for Myself . . . Just for Fun!

Here's a game to test your memory. Try it with a friend!

Put ten different things on a table. Have your friend look at them for ten seconds, then close her eyes or leave the room while you remove or hide one of the objects. When she returns, ask her which object is missing. Can she remember? If she does, she gets 1 point.

Then it's your turn. Your friend puts ten completely different things on the table. You look at them, leave the room, then come back and try to remember which one is missing. Each of you takes several turns, and the first player to get 5 points wins.

The more players for this game, the merrier!

Backward countdowns are a favorite activity of even very young children who are mastering their counting skills. Pretend that you're at Cape Canaveral for a shuttle launch. See how high a number your child can start with and correctly count down by ones to 0 ("30, 29, 28 . . . 3, 2, 1, 0, liftoff"). Children who can tell time can use the concept of counting down by ones to help get themselves, their brothers, and their sisters ready for school or other places (i.e., counting down the minutes before they have to leave, and acting accordingly—getting dressed, brushing their teeth, getting their backpacks ready, etc.).

My favorite number to take 1 away from is:

Here's a trick to help me remember how to subtract 1:

2. Me Two!

Zoo Animals

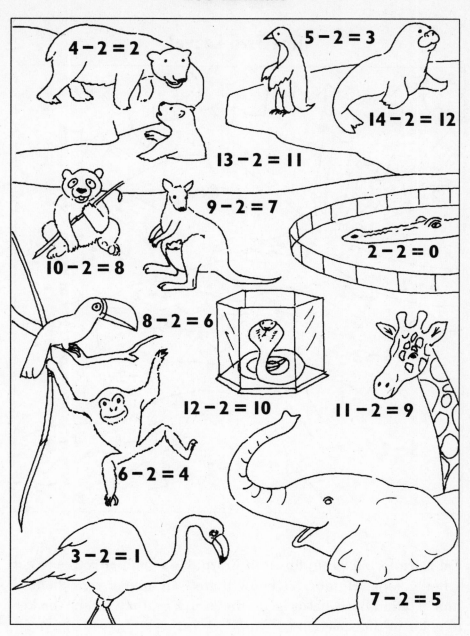

$4 - 2 = 2$

$5 - 2 = 3$

$14 - 2 = 12$

$13 - 2 = 11$

$9 - 2 = 7$

$10 - 2 = 8$

$2 - 2 = 0$

$8 - 2 = 6$

$12 - 2 = 10$

$11 - 2 = 9$

$6 - 2 = 4$

$3 - 2 = 1$

$7 - 2 = 5$

Two Activity:

Time Two Leave?

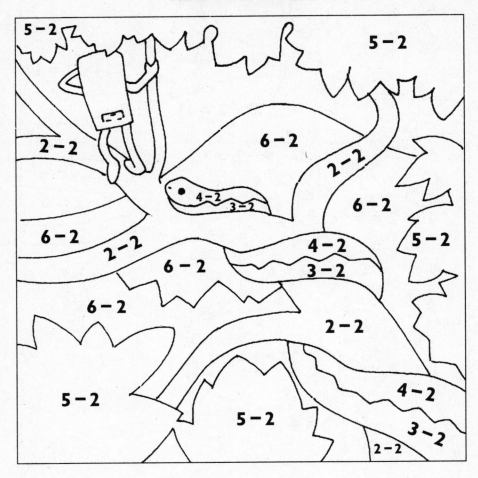

Cal is exploring a rain forest in Brazil. He's climbed a tree to get a better look. He doesn't know there's an animal right behind him. To find out what it is, do the exercises, then follow the key to color in the spaces.

0—black   3—green
1—red   4—brown
2—yellow

Team Sports

10 − 2 =

2 − 2 =

3 − 2 =

12 − 2 =

9 − 2 =

5 − 2 =

7 − 2 =

11 − 2 =

6 − 2 =

13 − 2 =

8 − 2 =

14 − 2 =

4 − 2 =

Become a Mathlete! Exercise with Callie!

Notes to Myself . . . Just for Fun

Lots of things come in pairs, or twos. How many can you think of? Here are some ideas to get you started:

Things people wear
Things people look through
Things people use

(You can also play this as a party game. Give your friends paper and pencils, and challenge them to come up with as many answers as they can in two minutes.)

Q: What day did the animals crawl, hop, fly, and slither onto Noah's Ark?

A: Twos-day (Tuesday)!

(1) Reinforce your child's grasp of odd and even numbers, either by writing them on paper or by manipulating objects. If 2 is subtracted from an odd number, what is the result? What if 2 is subtracted from an even number? Have the child see the pattern for him or herself. On paper, go all the way up to very large numbers (e.g., "Is 1,342 minus 2 an odd or even number? How about 745 minus 2?") until the child is confident. (2) In the animal kingdom, the differences between male and female can be striking. With a picture book or a videotape or a trip to the zoo, ask your child to identify some differences between male and female lions, bears, peacocks, hippos, gorillas, deer, penguins, etc. In which of these species do the male and female look different? In which do they look alike?

My favorite number to take 2 away from is:

Here's a trick to help me remember how to subtract 2:

3. Touch Your Toes, Three!

Zoo Animals

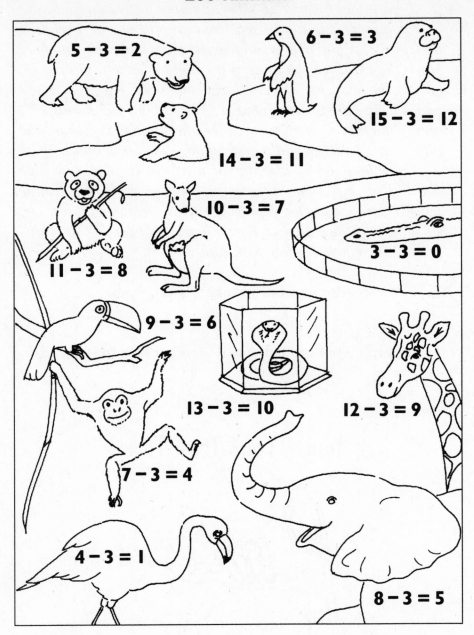

$5 - 3 = 2$

$6 - 3 = 3$

$15 - 3 = 12$

$14 - 3 = 11$

$10 - 3 = 7$

$3 - 3 = 0$

$11 - 3 = 8$

$9 - 3 = 6$

$13 - 3 = 10$

$12 - 3 = 9$

$7 - 3 = 4$

$4 - 3 = 1$

$8 - 3 = 5$

Three Activity:

Backward Bowling

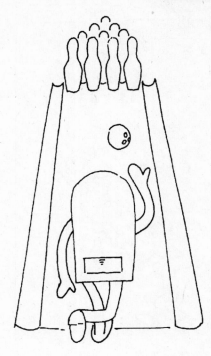

Remember that subtraction is reverse addition. So if you see a subtraction exercise of 5 – 3, you could either ask yourself, "Five take away 3 is what?" or "What number added to 3 would make 5?"

Cal has invented a new kind of bowling game. Players each start out with 10 points and subtract 1 point for each pin knocked down. The first player to reach 0 wins!

If Callie's ball knocks down 3 pins, what's her score? What if her next ball knocks down 2 pins?

Cal knocks down 6 pins. What's his score? What if his next ball knocks down 3 pins?

Three knocks down 4 pins with his first ball. What's his score? Then he knocks down 3 pins. What's his score now?

Who is closest to 0?

Team Sports

Become a Mathlete! Exercise with Callie!

Notes for Myself . . . Just for Fun!

Three is a very popular number in songs and stories. What fairy tales or nursery rhymes do you know that have the number 3 in them? Use the clues below to help you fill in the blanks.

1. These tiny animals couldn't see very well. They ran after the farmer's wife, who cut off their tails with a carving knife. Who are they? The three_____

2. A big bad wolf was chasing these chubby little guys. He kept blowing down their houses until one of them was smart enough to build a house of brick. Who are they? The three

_____

3. This furry family lived deep in the forest. One day a little girl with blond hair broke into their house, ate their food, and slept in their beds. Who are they? The three_____

(1) Many tales from the Brothers Grimm and others are built around triads such as three wishes, three gifts, or three brothers. Read some of these stories with your children and discuss them. (2) The next time you are with the kids in the car for a while, play a game in which someone names a category and the other players give three examples for that category. (For example, for the category Animals with Hooves, one player could say, "Sheep, horses, cows," and another player could say, "Camels, goats, oxen," and so on. Try useful categories such as Machines That Fly or Things I Can Do to Help the Earth, or silly categories such as Polka Dots I Have Known and Loved or Things I Wouldn't Want to Put in a Peanut Butter Sandwich.) Players score extra points for coming up with words that also begin with "t" or "th." Players who cannot think of an answer lose 3 points, except that no one can go below 0. The player with the highest point total at the end of play is the winner.

My favorite number to take 3 away from is:

Here's a trick to help me remember how to subtract 3:

4. Diving Four Fun

Notes to Myself . . . Just for Fun

Do you realize how many shapes have four sides? The most common are squares, rectangles, and diamonds. Draw one of each of these shapes below.

Think for a minute about these shapes. Are there things you see every day that are in the shape of a square? How about rectangles? How about diamonds? In the blanks below, write or draw four of these things.

_____   _____

_____   _____

There are other shapes with four sides, too. Ask your teacher or your parents if they can draw any of these other shapes. How are they different?

Zoo Animals

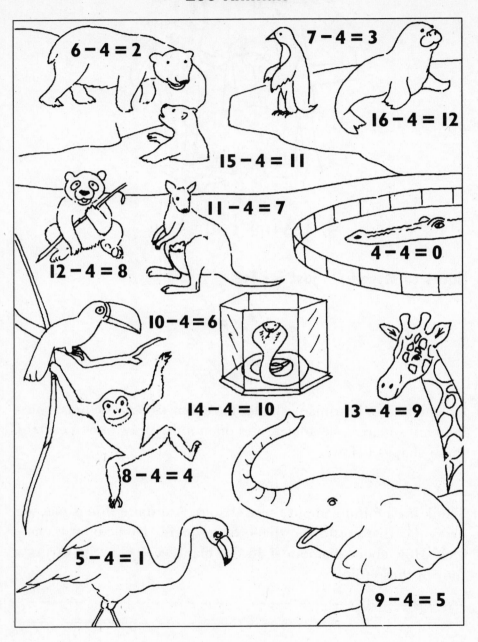

6 − 4 = 2

7 − 4 = 3

16 − 4 = 12

15 − 4 = 11

11 − 4 = 7

12 − 4 = 8

4 − 4 = 0

10 − 4 = 6

14 − 4 = 10

13 − 4 = 9

8 − 4 = 4

5 − 4 = 1

9 − 4 = 5

Team Sports

12 − 4 =

4 − 4 =

14 − 4 =

5 − 4 =

11 − 4 =

7 − 4 =

9 − 4 =

13 − 4 =

8 − 4 =

15 − 4 =

10 − 4 =

16 − 4 =

6 − 4 =

(1) Does your child know the four cardinal points of the compass? Review north, south, east, and west with him or her, then orient yourselves in the room you're in. Which wall is on the north side, which is on the south, and so on? If it's very early or very late in the day, note where the sun is in the sky, and help them understand that this is one way of determining east or west and thus the other directions as well. (2) Introduce your child to a compass, showing how the needle always points north. Extend this activity with magnets, a map, or a globe. (3) The books of L. Frank Baum have wonderful stories for reading with your children. In The Wizard of Oz, *for example, Dorothy is taken to the Land of Oz, which is divided into four quadrants: the purple land of Gillikin to the north, the blue land of the Munchkins to the east, the red land of Quadlings to the south, and the yellow land of Winkies to the west (with the Emerald City of Oz in the center).*

My favorite number to take 4 away from is:

Here's a trick to help me remember how to subtract 4:

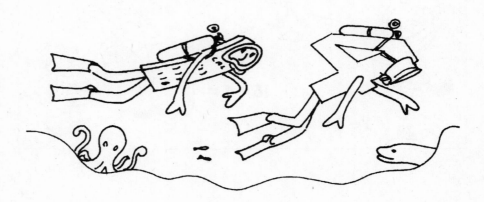

Become a Mathlete! Exercise with Cal!

5. Take a Hike, Five!

Become a Mathlete! Exercise with Callie!

Zoo Animals

Up and Down

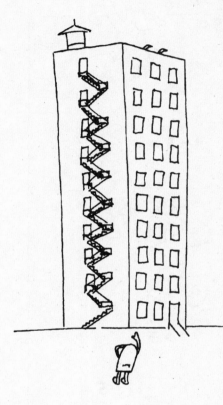

Pretend you're visiting a friend who lives in a tall apartment building. It has ten floors. If you start out on the first floor and take the elevator up five more floors, what floor do you get out on? (Hint: This is the same as 1 + 5.)

Let's say you get out of the elevator, then discover you went too high. Your friend actually lives on the floor below. You climb the stairs down one floor. What floor does she live on? You're now outside your friend's apartment, ringing her doorbell. She answers and says, "Let's pick up Sally and go to the park." Sally lives on the third floor. How many floors down do you have to go? Once you pick her up, how many more floors down to the first floor?

If you start out on the eighth floor, then go down five floors, where do you end up? Which floor do you get if you start out on the tenth one and go down five more?

Team Sports

13 − 5 =

5 − 5 =

15 − 5 =

6 − 5 =

12 − 5 =

8 − 5 =

10 − 5 =

14 − 5 =

9 − 5 =

16 − 5 =

11 − 5 =

17 − 5 =

7 − 5 =

Notes to Myself . . . Just for Fun

Do you remember the difference between odd and even numbers? Odd numbers include 1, 3, 5, 7, 9, and 11. Even numbers include 2, 4, 6, 8, 10, and 12.

What happens when you subtract 1 from an odd number? Try it with several odd numbers and see. What happens when you subtract 1 from an even number? Try it with several of those.

Did you notice that when you subtract 1 from an odd number, the answer is always an even number? And did you also find out that if you take away 1 from an even number, the answer is always an odd number? You are smart! But what if you subtract 2? Try it and see. First take away 2 from an odd number, and see what happens. Now take away 2 from an even number. Did you discover that when you subtract 2 from an odd number, the answer is always another odd number? And if you take 2 away from an even number, the answer is always another even number.

Is 5 an even or an odd number? If you subtract 5 from an even number, what kind of number do you think the answer will be? What if you subtract 5 from an odd number? (Hint: Try subtracting 5 from different even numbers, such as 6, 8, and 10. Then subtract it from different odd numbers, such as 7, 9, and 11. What pattern do you find?)

Encourage your child to experiment with several of these numbers. Help them find patterns that will help them in subtracting.

Play the "Nickel Game" with your child. Both of you start out with fifty cents' worth of nickels (or counters representing five cents each) and come up with your own list of "forbidden activities" to avoid doing for the next one or five minutes. Examples of such activities could be scratch-

ing one's nose, saying the word "like," crossing one's legs, giggling, etc. Whoever does the forbidden activity is "fined" five cents and gives it to your opponent. The person with the most nickels at the end of play is the winner.

 My favorite number to take 5 away from is:

Here's a trick to help me remember how to subtract 5:

6. Mix It Up with Six

Did you remember that subtraction is really reverse addition? To solve an exercise such as 10 − 6, you could either ask yourself, "Ten take away 6 is what?" or "What number added to 6 would make 10?"

Zoo Animals

8 − 6 = 2

9 − 6 = 3

18 − 6 = 12

17 − 6 = 11

13 − 6 = 7

6 − 6 = 0

14 − 6 = 8

12 − 6 = 6

16 − 6 = 10

15 − 6 = 9

10 − 6 = 4

7 − 6 = 1

11 − 6 = 5

Six Activity:

Crack the Code

That globe-trotter Cal! Now he's in Egypt, searching for clues about ancient civilizations. Help him read the secret symbols written on the tomb. Do each exercise below, then put each letter on the blank that matches its answer.

12 − 6 = N

7 − 6 = A

10 − 6 = M

8 − 6 = T

9 − 6 = I

16 − 6 = Y

18 − 6 = W

15 − 6 = U

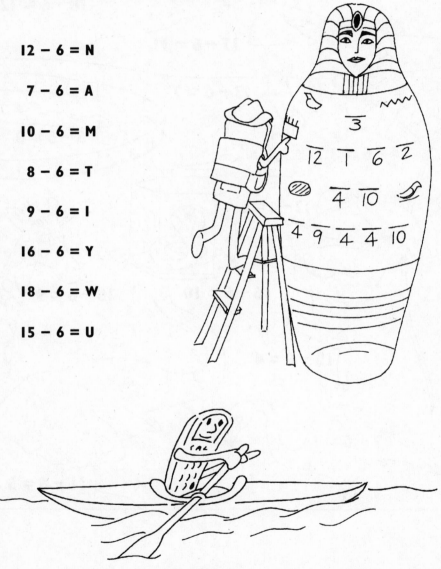

Become a Mathlete! Exercise with Cal!

Team Sports

14 − 6 =

6 − 6 =

7 − 6 =

16 − 6 =

13 − 6 =

9 − 6 =

11 − 6 =

15 − 6 =

10 − 6 =

17 − 6 =

12 − 6 =

18 − 6 =

8 − 6 =

Notes for Myself . . . Just for Fun!

Have you ever been to a museum? Most museums have large collections of interesting things: dinosaur skeletons, fossils, and art and tools from other places and times.

If you or your friends like to collect things, maybe you can open your own "museum" (or ask your teacher if your class can start one). Maybe Joel has a bunch of shells he's really proud of . . . Meredith gathers bird feathers and cocoons . . . Gabriel collects pictures of submarines.

How many things does each person have? How many different ways can these things be sorted? (For example, of Joel's shells, how many are round? flat? shiny? smooth? rough? white? brown?)

(1) Have your child imagine that your house is a museum and that you have collections of six of all kinds of things of different sizes. Which collection could fit in the palm of someone's hand? Which could fit in a box? a room? Which might take up the entire house? (2) Ask your child to name a quick way of making a six-pack of water or juice disappear. When would it be especially appropriate to use such a method or methods? (3) Snowflakes have six sides, and no two flakes are exactly alike. What are your child's favorite things to do in the snow? See who can "capture" a snowflake

and "hold on to it" the longest. How could someone extend the "life" of a snowflake?

 My favorite number to take 6 away from is:

Here's a trick to help me remember how to subtract 6:

7. Seven's Heaven

Become a Mathlete! Exercise with Cal!

Zoo Animals

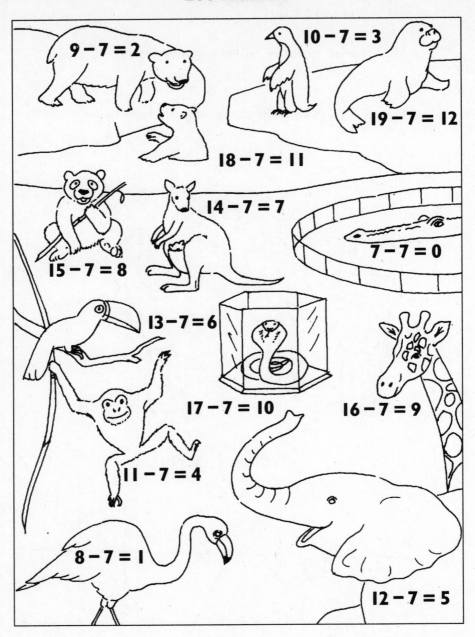

9 − 7 = 2

10 − 7 = 3

19 − 7 = 12

18 − 7 = 11

14 − 7 = 7

15 − 7 = 8

7 − 7 = 0

13 − 7 = 6

17 − 7 = 10

16 − 7 = 9

11 − 7 = 4

8 − 7 = 1

12 − 7 = 5

Team Sports

Notes to Myself . . . Just for Fun

Just about everybody knows the story of Snow White and the Seven Dwarfs. If you really wanted to have some fun with the story, what would you do to make each of Snow White's friends as different as could be? Would you have each of the dwarfs doing different things, like surfing or hot-air ballooning? Could they be on a football team wearing different numbers on their jerseys? (How silly! The Dallas Dwarfs or Snow White's Seahawks.) How would you draw them differently? What is each one's favorite color, song, or food?

 What do these exercises have in common?

10 − 3
12 − 5
17 − 10
9 − 2
11 − 4
8 − 1
77 − 70

How familiar is your child with calendars? He or she can help you count down by sevens (for seven days in a week) to a particular date, such as someone's birthday, or a holiday, or the day you go on vacation, or next Sunday. Simply show your child what day it is today on the calendar, and which day is seven days away. Then he or she can mark off each day, or cut a link out of a paper chain, etc. After this you can ask your child how many weeks he or she seems to notice in a month.

My favorite number to take 7 away from is:

Here's a trick to help me remember how to subtract 7:

8. Make a Date with Eight

Zoo Animals

$10 - 8 = 2$

$11 - 8 = 3$

$20 - 8 = 12$

$19 - 8 = 11$

$15 - 8 = 7$

$16 - 8 = 8$

$8 - 8 = 0$

$14 - 8 = 6$

$18 - 8 = 10$

$17 - 8 = 9$

$12 - 8 = 4$

$9 - 8 = 1$

$13 - 8 = 5$

Team Sports

Become a Mathlete! Exercise with Cal!

Notes to Myself . . . Just for Fun

Numbers are important in the animal kingdom because they give us one way of telling certain kinds of animals apart. For instance, did you know that ants, bees, and spiders are not just "bugs"? They're completely different critters. Spiders—like daddy longlegs and tarantulas—have eight legs, and scientists call them arachnids. But ants and bees have six legs, and we know them as insects.

Count the number of legs on the ladybug. Is it an insect or an arachnid? How about the fly? Ask your teacher or parents if they know other kinds of arachnids besides spiders.

 (1) Help your child learn about different kinds of bugs and arachnids. If a person wanted to keep some in a small cage for a while, what would the critters need? Make sure your child releases the animal before the day is over. Also make sure your child knows which animals are dangerous and to be avoided. (2) Did you know that there are more species of insects on this planet than any other life-form?

My favorite number to take 8 away from is:

Here's a trick to help me remember how to subtract 8:

9. On Cloud Nine

Become a Mathlete! Exercise with Cal!

Zoo Animals

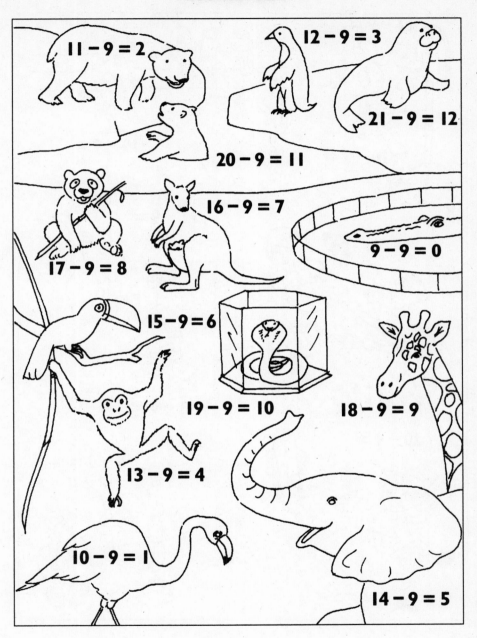

$11 - 9 = 2$

$12 - 9 = 3$

$21 - 9 = 12$

$20 - 9 = 11$

$16 - 9 = 7$

$17 - 9 = 8$

$9 - 9 = 0$

$15 - 9 = 6$

$19 - 9 = 10$

$18 - 9 = 9$

$13 - 9 = 4$

$10 - 9 = 1$

$14 - 9 = 5$

Team Sports

Notes to Myself . . . Just for Fun

Think of real-life situations where subtraction takes place. There are lots of ways that things can be taken away, or "disappear," or be changed into something else. Choose one of the following activities and make up a story about it. Be sure to include some kind of subtraction in your story (you can do this by using words like "take away" or "less" or "minus").

1. You and your family go on a picnic.
2. "Marvin the Magician" performs at your school.
3. A firefighter climbs up a ladder, rescues some people, then climbs back down the ladder.

The process of descent can be likened to subtraction. If an adult is at the top of an extension ladder and goes down nine steps, that would be one example, for the person starts at, say, step ten and goes down to step one. Ask your child for other examples, such as going downstairs nine steps or nine flights, or going down nine floors in an elevator, or driving nine kilometers down a mountain road. What do scuba divers do? Submarines? Since climbing down is like subtraction, what is addition like?

 My favorite number to take 9 away from is:

Here's a trick to help me remember how to subtract 9:

10. Strike It Rich, Ten

Ten Activity:

Ring Around the Collar

Ten enjoys bowling almost as much as surfing—after all, it's a game that does use ten pins! He and Cal have played several games, but the scoring machine is acting funny. For a lot of the scores, it's substituting equations. Can you look at each pair of scores on page 141 and "put a ring" (circle) around the one that's bigger?

10 − 10, 10 − 1
3, 12 − 10
10 − 8, 11 − 10
10, 11
10 + 1, 10
10, number of letters in "ten"

Bowling is a game that has been around in one way or another, in different places, for at least seven hundred years. Have you ever heard the tale of Rip Van Winkle? As the story goes, Rip was a man who lived during colonial times in what is now New York. One day he wandered into the Catskill Mountains and played a type of bowling game, called ninepins, with some very strange folk he met there. Then he slept for twenty years! When he woke up, the world he knew had changed.

Take your children to the library and introduce them to the stories of Washington Irving. Another popular one besides Rip Van Winkle is "The Legend of Sleepy Hollow."

Become a Mathlete! Exercise with Callie!

Zoo Animals

12 − 10 = 2

13 − 10 = 3

22 − 10 = 12

21 − 10 = 11

17 − 10 = 7

10 − 10 = 0

18 − 10 = 8

16 − 10 = 6

20 − 10 = 10

19 − 10 = 9

14 − 10 = 4

11 − 10 = 1

15 − 10 = 5

Team Sports

18 − 10 =

10 − 10 =

20 − 10 =

11 − 10 =

17 − 10 =

13 − 10 =

15 − 10 =

19 − 10 =

14 − 10 =

21 − 10 =

16 − 10 =

22 − 10 =

12 − 10 =

Subtracting by 10s is a snap! Simply cross out the 0, do the exercise, then write the 0 back into the answer.

For example, if you want to take 10 away from 20, first cross out the 0s so it looks like 2 – 1. The answer is 1. Add the 0 back to the 1, and you have your answer: 10.

$$\begin{array}{r} 20 \\ -\ 10 \\ \hline \end{array}$$

$$\begin{array}{r} 2 \\ -\ 1 \\ \hline 1 \end{array}$$

$$10$$

Use this trick to do the following:

$$\begin{array}{r} 40 \\ -\ 30 \\ \hline \end{array}$$

$$\begin{array}{r} 40 \\ -\ 20 \\ \hline \end{array}$$

$$\begin{array}{r} 50 \\ -\ 10 \\ \hline \end{array}$$

$$\begin{array}{r} 80 \\ -\ 50 \\ \hline \end{array}$$

$$\begin{array}{r} 100 \\ -\ 50 \\ \hline \end{array}$$

Remember to put the 0 back when you're done!

(1) Do you have a wall chart to see how much your child has grown in the past year? Show the mark you made for your child's height last year, or six months ago, and compare it with today's. How much of a difference is there? Ask your child to subtract ten inches (or other increments) and see if there was a mark for a past height that corresponds roughly. How long ago was it? (2) Do the same thing on a bathroom scale. Do you remember when your child weighed ten pounds less? How long ago was it? (3) Another activity to do on the scale is to have everybody take turns weighing just themselves, then holding heavy objects (e.g., the pet dog) and weighing themselves again, then subtracting to find out how much the object weighs. (4) Other measuring devices with which you and your child can practice subtracting 10 include a kitchen scale, a postage meter, and measuring cups.

My favorite number to take 10 away from is:

Here's a trick to help me remember how to subtract 10:

11. Eleven's Excellent Adventure

Become a Mathlete! Exercise with Cal!

Zoo Animals

13 − 11 = 2

14 − 11 = 3

23 − 11 = 12

22 − 11 = 11

18 − 11 = 7

19 − 11 = 8

11 − 11 = 0

17 − 11 = 6

21 − 11 = 10

20 − 11 = 9

15 − 11 = 4

12 − 11 = 1

16 − 11 = 5

Eleven Activity: Crazy Classroom

Can you circle all the things that are wrong with this picture?

Now look at the exercises on the chalkboard. Some of them have crazy answers. Draw an "X" through the wrong answers, then write in the correct ones.

Team Sports

19 − 11 =

11 − 11 =

21 − 11 =

12 − 11 =

18 − 11 =

14 − 11 =

16 − 11 =

20 − 11 =

15 − 11 =

22 − 11 =

17 − 11 =

23 − 11 =

13 − 11 =

Notes to Myself . . . Just for Fun

Different games are played with different types of scoring. Use the following clues to guess which game is being described:

1. Each goal in this game is worth 1 point. If a player scores three goals in one game, that's called a "hat trick"—it's quite a nice accomplishment!

2. Most points are scored 1 at a time, but if you're really striking you get a bonus!

3. The crowd is on its feet when the first 6 points are scored, but if the team can get that additional 1, it's a real kick!

4. Most points are earned 15 at a time. Fans of this game absolutely love it!

Does your child have a favorite team sport? How about soccer, baseball, football, hockey, basketball, volleyball? How many players on a team? How are points scored? How many periods, innings, etc. in a game? What other numbers apply to that sport?

My favorite number to take 11 away from is:

Here's a trick to help me remember how to subtract 11:

12. Dozen't This Look Like Fun?

Become a Mathlete! Exercise with Callie!

Zoo Animals

14 − 12 = 2

15 − 12 = 3

24 − 12 = 12

23 − 12 = 11

19 − 12 = 7

12 − 12 = 0

20 − 12 = 8

18 − 12 = 6

22 − 12 = 10

21 − 12 = 9

16 − 12 = 4

13 − 12 = 1

17 − 12 = 5

Twelve Activity:
Landlubbers Ahoy

Avast, matey! Cal has discovered an old pirate's map on the beach. But the wind and the waves have washed away part of the message. Can you help Cal figure out what it says?

First fill in the missing numbers in the exercises. Then, if the answer matches a numbered blank below, transfer its corresponding letter to that blank. (Not all the letters will be used.)

One exercise has been done for you. (Hint: There are different ways to solve these. One way is to ask, "Three added to what equals 8?" Another way is to ask, "Eight minus what equals 3?")

12 − ___ = 0	N		16 − 12 = ___	V
20 − 12 = ___	T		23 − 12 = ___	F
16 − ___ = 3	P		13 − 12 = ___	H
12 − 12 = ___	U		12 − ___ = 3	R
7 − ___ = 2	E		19 − 12 = ___	S

Team Sports

Notes to Myself . . . Just for Fun

Did you ever notice that a lot of things having to do with time come in twelves? How about the hours on the face of a clock? or the number of months in a year? or the signs in the zodiac? What do you know about other types of calendars (such as the Chinese or Jewish calendar)? Do you know anyone who celebrates a different new year from the one on January 1?

Q: If it takes two kids two hours to wash two cars, how many hours does it take for one kid to wash one car?

A: Two hours.

(1) Reinforce your child's time-telling ability by showing him or her different times on a clock or watch. What activities take place at different times? At 7:00 A.M., for example, is a person more likely to be waking up or going to bed? At 6:00 P.M. is a person more likely to be eating dinner or taking the bus to school? (2) Ask your child if everyone around the world is doing the same things at the same

time as you, then discuss different time zones. Rotate a globe as you do so to convey both the earth's relation to the sun and how people in a particular spot on the globe are affected. (3) How many different clocks are there in your house? Have your child identify them all. Don't forget devices such as microwave ovens, VCRs, answering machines, and computers.

 My favorite number to take 12 away from is:

Here's a trick to help me remember how to subtract 12:

Subtraction Review

We've really come a long way, haven't we? Now you're an expert at subtraction! Whenever you take something away from something else, then see how many are left, you are subtracting. When you subtract, the answer is called the *difference*. For example, when you take away 2 from 3, the difference is 1 — this is written as $3 - 2 = 1$.

Subtraction is the opposite of addition, and you can think of it as *reverse addition*. So another way of thinking about subtraction is to ask yourself, "What plus 2 is 3?" For some people, this is easier than asking themselves, "Three take away 2 is what?" In either case, the answer is 1. Just do what works best for you!

Subtraction is a snap in lots of ways. Remember how easy it was to add 0 to any number? Subtracting 0 from any number is just as easy. The answer is always the same as what you started out with. That's because when you take away 0, you're taking away nothing! Subtracting 1 from any number is also a breeze. The difference is always the number that comes right before the number you started out with. For example, the number just before 5 is 4 — and what's $5 - 1$? That's right, it's 4!

Something else that's fun to know about subtracting 1 is this: If you take away 1 from an *even* number (like 2 or 4), the difference is always an odd number ($2 - 1 = 1$; and $4 - 1 = 3$). If you take away 1 from an *odd* number (like 3 or 5), the difference is always an even number ($3 - 1 = 2$; and $5 - 1 = 4$).

A shortcut to help you subtract 2 from any number is this: If you take away 2 from an even number, the difference is always an even number. If you take away 2 from an odd number, the difference is always an odd number.

Subtracting 10 from big numbers that end in 0 is also a breeze. Simply cross out or take off the 0s (only for a little while, though), then do the subtraction with the numbers that remain.

Then write a 0 back in as the last part of the answer. For example:

$$
\begin{array}{r}
50 \\
-\ 10 \\
\end{array}
$$

Get rid of the 0s for now, then subtract.

$$
\begin{array}{r}
5 \\
-\ 1 \\
\hline
4 \\
\end{array}
$$

Write a 0 after the 4 to get your answer:

40

Aren't you smart!

5

Growing by Leaps and Bounds

Multiplication with the Human Calculator

Multiplication is *fast addition*. When we want to add a group of numbers that are all the same, we just see how many there are and multiply by that number. Let's look at the same problem, done two different ways.

Addition

```
  I
  I
  I
 +I
  4
```

Multiplication

```
  I
 ×4
  4
```

Adding 1 to itself four times gives you the same answer as multiplying 1 by 4, but which way is quicker? That's right—once you know multiplication, you'll find that it gives you answers much faster than addition. That's because multiplication is a way of grouping things together. Let's look at the above problem again, only this time with squares:

$$I + I + I + I = 4 \ = \ \square\ \square\ \square\ \square$$

$$= \ \boxed{\ \ \ \ \ \ \ } \ = \ I \times 4 = 4$$

They both say the same thing, except that the top one uses addition and the bottom one uses multiplication. Here's another example:

$$2 + 2 + 2 + 2 = 8 \quad =$$

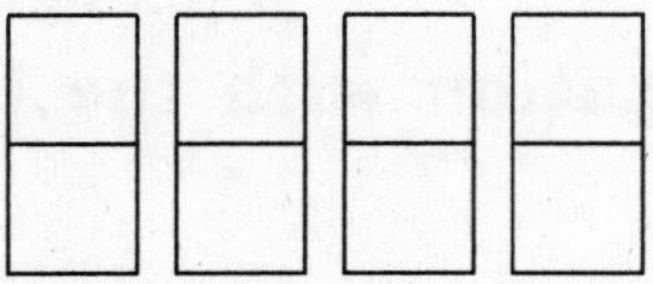

$$=$$

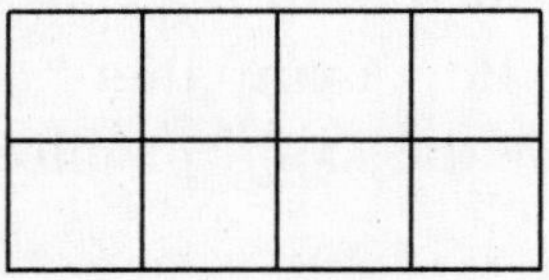

$$= 2 \times 4 = 8$$

Again, both are the same, except that once you know multiplication, it lets you count even faster than addition.

In multiplication, the answer is called a *product.* In the multiplication equation $2 \times 3 = 6$, the *product* is 6. Circle the products in the following:

$$\begin{array}{r} 2 \\ \times 4 \\ \hline 8 \end{array} \qquad \begin{array}{r} 4 \\ \times 1 \\ \hline 4 \end{array} \qquad \begin{array}{r} 6 \\ \times\ 5 \\ \hline 30 \end{array} \qquad \begin{array}{r} 12 \\ \times\ 2 \\ \hline 24 \end{array}$$

That's right! The products are 8, 4, 30, and 24.

0. Zero the Hero

Multiplication is when Zero really gets to show his stuff: Any number multiplied by 0 actually *becomes* 0! Zero has this power over any number, no matter how big it is. In fact, you could say that this is when Zero truly becomes a "Super-Zero"! For example:

$1 \times 0 = 0$
$2 \times 0 = 0$
$3 \times 0 = 0$
$10 \times 0 = 0$
$xyz \times 0 = 0$
$4{,}567{,}843{,}000 \times 0 = 0$

Forest Animals

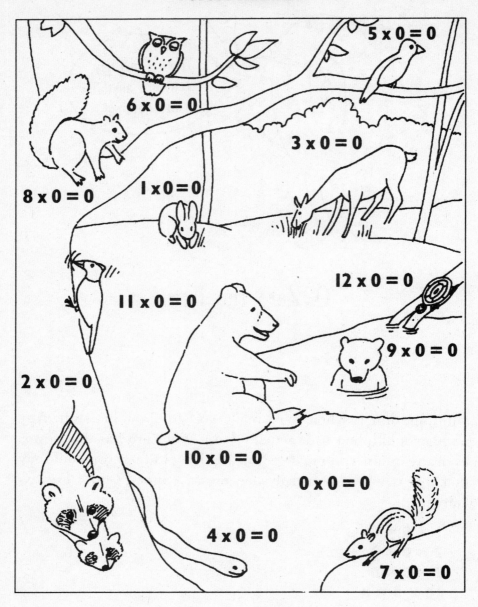

$5 \times 0 = 0$

$6 \times 0 = 0$

$3 \times 0 = 0$

$8 \times 0 = 0$

$1 \times 0 = 0$

$12 \times 0 = 0$

$11 \times 0 = 0$

$9 \times 0 = 0$

$2 \times 0 = 0$

$10 \times 0 = 0$

$0 \times 0 = 0$

$4 \times 0 = 0$

$7 \times 0 = 0$

Zero Activity:

The Winning Way

Zero is competing in a karate tournament. Help him by doing the exercises in the maze below, then drawing a path that leads him to the trophy.

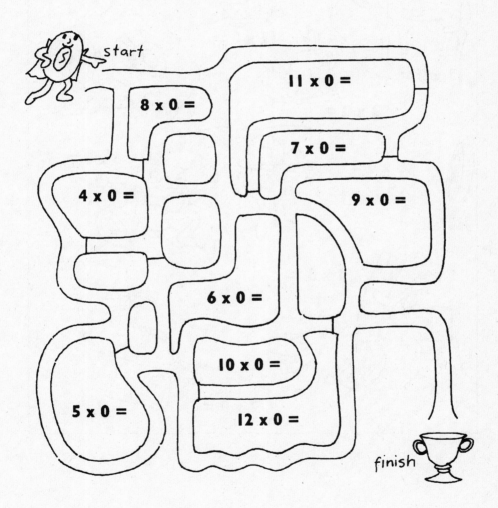

start

11 x 0 =

8 x 0 =

7 x 0 =

4 x 0 =

9 x 0 =

6 x 0 =

10 x 0 =

5 x 0 =

12 x 0 =

finish

Camp Activity

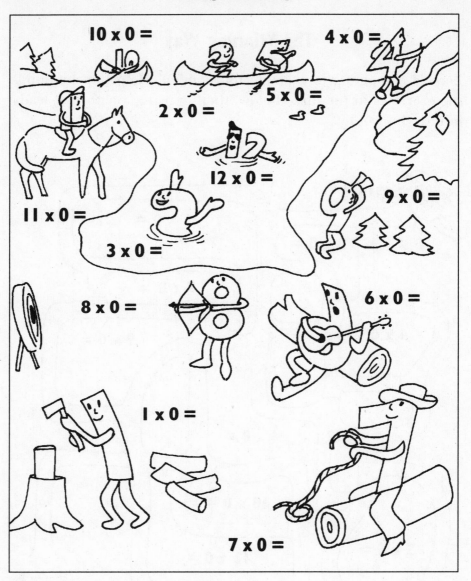

10 x 0 =

4 x 0 =

5 x 0 =

2 x 0 =

12 x 0 =

11 x 0 =

9 x 0 =

3 x 0 =

8 x 0 =

6 x 0 =

1 x 0 =

7 x 0 =

Become a Mathlete! Exercise with Cal!

Notes for Myself . . . Just for Fun

Did you know that *karate* is a Japanese word meaning "empty hand"? It's a highly effective form of self-defense that uses no weapons. If you think about it, that's kind of what 0 does in multiplication. Zero can be a "nothing" sometimes, but when he's multiplying, watch out—he can become quite a powerful weapon himself!

The sport of karate is very popular with boys and girls. It can be a great way to build strength, discipline, and self-confidence. Different colors of belts are earned by karate students as they add to their skills. The colors depend on each karate school. For example, one school awards the following colors of belts, from beginning to advanced: white, blue, green, purple, and brown.

Color the belts that Cal, Callie, and their friends are wearing on the next page. Use any colors you want. Which color do you want to stand for the beginning level? Which colors are more advanced? Which color is worn only by experts?

"A Zero by Any Other Name." One of the great things about 0 is that it's the same as nothing, and there are lots of funny and crazy ways to describe nothing. For example, if you say "the number of walruses that live in trees," that's the same as saying "nothing" or 0 (unless someone *does* find a walrus that lives in a tree, of course!). Another example of nothing or 0 is "fleas that are bigger than the earth."

Show your teacher and friends how smart you are by coming up with as many wacky ways of describing 0 as you can. Here are some examples to get you started:

0 × 3 = the number of sharks that dance
12 × 0 = cars that run on peanut butter
6 × 0 = wings on a giraffe

Explore with your child various ways in which things can be erased or neutralized. Point out, for example, that the effect of erasing a chalkboard is like multiplying by 0. How can the child do the same thing with a pencil and paper? with white paint? For older children who know appropriate behavior, ask how someone would "neutralize" an audiotape with a tape player or answering machine. How would a videotape be erased in a VCR? Can a computer diskette be initialized?

My favorite number to multiply 0 by is:

Here's what I like best about multiplying by 0:

1. On Target with One

As you have found out, any number multiplied by 1 stays the same. For example:

1 × 1 = 1
2 × 1 = 2
3 × 1 = 3
10 × 1 = 10
xyz × 1 = *xyz*

Forest Animals

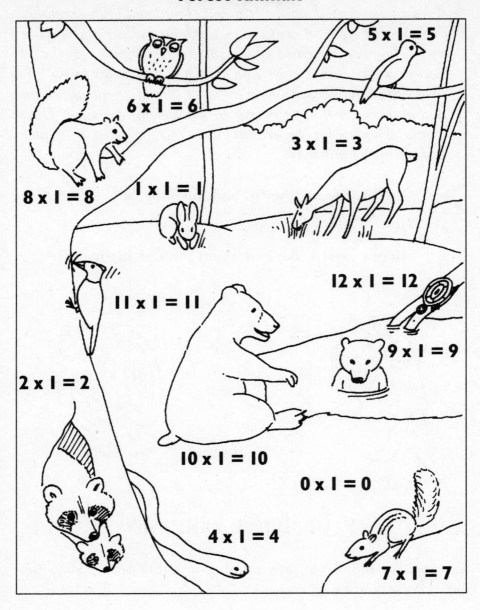

$5 \times 1 = 5$

$6 \times 1 = 6$

$3 \times 1 = 3$

$8 \times 1 = 8$

$1 \times 1 = 1$

$12 \times 1 = 12$

$11 \times 1 = 11$

$9 \times 1 = 9$

$2 \times 1 = 2$

$10 \times 1 = 10$

$0 \times 1 = 0$

$4 \times 1 = 4$

$7 \times 1 = 7$

Go Fly a Kite!

It's a beautiful spring day, and Cal and Callie have gone to the park with some of their friends. To find out what's on Callie's kite, look at the equations below. Circle the letter corresponding to whether the equation is true or false. Write the circled letters down, then unscramble them.

Multiplication Equation	True	False
$1 \times 1 = 2$	A	E
$1 \times 8 = 8$	R	C
$4 \times 1 = 1$	Y	Z
$1 \times 1 = 1$	B	D
$10 \times 1 = 10$	E	P
$2 \times 1 = 3$	G	E

Camp Activity

10 x 1 =

4 x 1 =

5 x 1 =

2 x 1 =

12 x 1 =

11 x 1 =

9 x 1 =

3 x 1 =

8 x 1 =

6 x 1 =

1 x 1 =

7 x 1 =

Become a Mathlete! Exercise with Callie!

Notes for Myself . . . Just for Fun

Q: What were the first flying machines?

A: Kites! Did you know that more than three thousand years ago, boys and girls in China flew kites made out of silk? Today, kites can take many shapes. The most basic kind is flat and shaped like a diamond. A box kite is built around a frame made of squares, rectangles, and other shapes. Some kites are very long, made of pieces strung together in a row. This type of kite can be painted to look like a snake, dragon, or other animal.*

(1) Introduce your child to quantities that equal each other, such as in baking or measuring. For example, take a measuring cup and show with water how 8 ounces are the same as 1 cup, and how 2 cups equal 1 pint. (Use metric quantities.) Or take a yardstick and show how 12 inches equal 1 foot, and 3 feet equal 1 yard. (2) Have your child help you prepare a snack or meal using a recipe with

*"Kites and Gliders," *The Random House Children's Encyclopedia* (New York: Random House, 1991), p. 305.

some of the quantities he or she has discovered with the measuring cup.

 My favorite number to multiply 1 by is:

Here's what I like best about multiplying by 1:

2. Two in a Canoe

Any number multiplied by 2 doubles! In other words, just add the number to itself. For example, 3 multiplied by 2 is the same as adding 3 to 3 (to get 6). Whenever you multiply by 2, you wind up with twice as many as you started out with.

3 × 2 = 6

Forest Animals

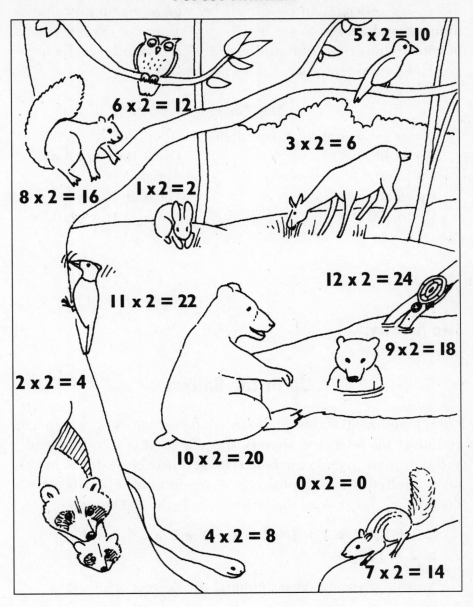

5 x 2 = 10

6 x 2 = 12

3 x 2 = 6

8 x 2 = 16

1 x 2 = 2

11 x 2 = 22

12 x 2 = 24

9 x 2 = 18

2 x 2 = 4

10 x 2 = 20

0 x 2 = 0

4 x 2 = 8

7 x 2 = 14

Have you ever counted by 2s? You know—2, 4, 6, 8, 10, 12, and so on. It's called "skip counting" because you skip every other number, and it's fun to see how high you can go. Try it with one of your friends right now and see who can count by 2s the highest. Here's a chart to check your progress.

Skip Counting

If you can count by 2s up to 20	Good job!
Up to 40	Great skip counting!
Up to 60	Terrific!
Up to 80	What an expert!
Up to 100	You're a true Mathlete!

Two Activity:

Up in the Rafters

Cal is helping out an old friend by taking inventory of (that means counting) his feathered friends. But it's starting to rain outside, and Cal wants to go home before he gets drenched. If Cal has already spotted two each of twelve different kinds of birds, what's the quickest way he can total them? Circle the right answer.

a. 2 + 2 + 2 + 2 + 2 + 2 + 2 + 2 + 2 + 2 + 2 + 2
b. 2 × 12

Now do these exercises, circling the correct answers.

$$2 \times 1 =$$

a. **2** b. **1** c. **3**

$$2 \times 3 =$$

a. **5** b. **6** c. **8**

$$2 \times 6 =$$

a. **12**   b. **8**   c. **9**

$$2 \times 5 =$$

a. **15**   b. **7**   c. **10**

$$2 \times 7 =$$

a. **14**   b. **10**   c. **9**

$$2 \times 9 =$$

a. **16**   b. **12**   c. **18**

Q: There are four kinds of birds that are a lot like the number 2. What are they?

A: toucans ("two" cans), parrots ("pair"-ots), cockatoos (cock-a-"twos"), and parakeets ("pair"-a-keets)

Become a Mathlete! Exercise with Cal!

Camp Activity

10 x 2 =

4 x 2 =

5 x 2 =

2 x 2 =

12 x 2 =

11 x 2 =

9 x 2 =

3 x 2 =

8 x 2 =

6 x 2 =

1 x 2 =

7 x 2 =

Multiplication and addition have something in common: It doesn't matter in what order you add or multiply the numbers; the answer is always the same. For example, if you're adding 3 to 2, you can start with either number because the sum will always be 5. And if you're multiplying 3 by 2, you can start with either number because the product will always be 6.

Have you also noticed that whenever you multiply by 2, the answer is always even? Try it and see!

Which of the following does *not* equal 8?

a. 2×4 b. $4 + 4$ c. $2 + 2 + 2$ d. 4×2 e. $2 + 2 + 2 + 2$

Notes for Myself . . . Just for Fun

Do you know anyone who is a twin? Maybe you're lucky enough to be one yourself! There are two different types of twins. *Identical* twins look very much alike and will either be two boys or two girls. People who don't know them well may not be able to tell them apart! *Fraternal* twins look different from each other. They can either be two boys, two girls, or a boy and a girl.

There are lots of stories, plays, and movies about twins (or about two people who look so much like each other that they *could* be twins!). A great English writer named William Shakespeare had twins in some of his plays; one of these is called *Twelfth Night*. Another well-known story is about two boys from different families who look enough alike to fool people. It's called *The Prince and the Pauper*, and it was written by Mark Twain. Ask your parents or teacher to read these with you.

Q: Raymond and Zachary are twins, but they don't have the same birthday. How can that be?

A: There are several possible answers. One is that, although Raymond and Zachary are twins, they are not *each other's* twins (they come from different families). Another possible answer is that they are indeed each other's twins, but that one was born just before midnight, and the other a few minutes after midnight, giving them separate birth dates.

 Next time you are shopping with your child, locate items that are available in different quantities, and show him or her the unit price for an item (e.g., ten cents per ounce), which should be posted on the shelf. Then look at a size that is double the original size. With your child, multiply by 2 to determine what would be an expected price for the double quantity, then check what it actually is. In many cases (but not all!) the unit price is lower for a larger quantity (e.g., eight cents per ounce). Discuss with your child how buying larger sizes is often more economical, and how the best thing to do is always to check and compare.

My favorite number to multiply 2 by is:

Here's what I like best about multiplying by 2:

3. Three Up a Tree

Become a Mathlete! Exercise with Cal!

Forest Animals

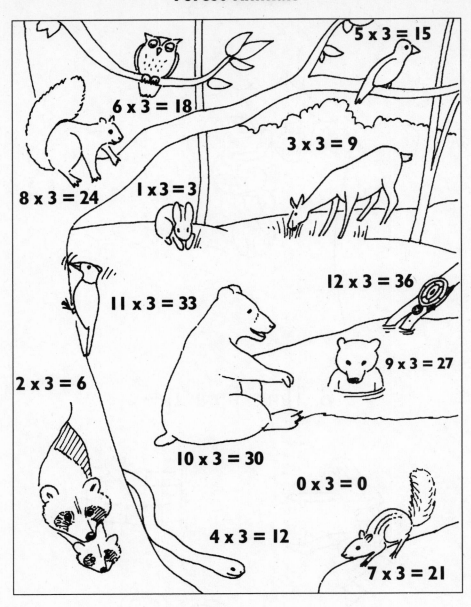

5 x 3 = 15

6 x 3 = 18

3 x 3 = 9

8 x 3 = 24

1 x 3 = 3

11 x 3 = 33

12 x 3 = 36

9 x 3 = 27

2 x 3 = 6

10 x 3 = 30

0 x 3 = 0

4 x 3 = 12

7 x 3 = 21

> Remember that multiplication is *fast addition*!

Did you enjoy counting by 2s earlier? Skip counting is fun for 3s, too! Try it and see: 3, 6, 9, 12 . . . You can turn this into a game with one of your friends. Start counting by 3s and keep going until you miss; then your friend takes a turn and keeps going until he or she misses. Then it's your turn again. Keep practicing to see how far you can go.

 If you want to keep track as you skip count by 3s, here's a chart you can fill in. The higher you get, the more you can impress your friends!

3	6		12		18			27
		36		42			51	
57			66			75		81
		90			99			108

Camp Activity

10 x 3 =

4 x 3 =

5 x 3 =

2 x 3 =

12 x 3 =

9 x 3 =

11 x 3 =

3 x 3 =

8 x 3 =

6 x 3 =

1 x 3 =

7 x 3 =

Grouping tricks let you use what you already know about certain numbers to make it easier to multiply by other numbers. Whenever Cal uses a grouping trick, it reminds him of a diving trip he took in the Pacific Ocean. Everywhere he swam, he was followed by fat, funny, friendly fish called *groupers*. Whenever you see the little groupers pop up, you'll know that it's a good time to use your grouping tricks.

Let's say you're already comfortable with multiplying by 1 and 2, but you're a little unsure about 3. You want to multiply 5 by 3. No problem! Just "rearrange" the 3 by doing this:

First, remember that 3 is really 2 + 1. So, multiply 5 by 2:

5 × 2 = 10

Then multiply 5 by 1:

5 × 1 = 5

Now add the two sums together:

10 + 5 = 15

The answer, 15, is the same as what you would have gotten if you had multiplied 5 by 3.

Let's try this with another equation: 6 × 3.

First, remember that 3 is really 2 + 1. So, multiply 6 by 2:

6 × 2 = 12

Then multiply 6 by 1:

6 × 1 = 6

Now add the two sums together:

12 + 6 = 18

Notes for Myself . . . Just for Fun

No matter what age you are, it's very important to eat nutritious foods that help you grow and stay healthy. The Food Pyramid is an easy guideline to follow, as shown here. Each day, eat lots of *carbohydrates* (whole grains, pasta, bread), *vegetables* (broccoli, carrots, cabbage), and *fruits* (apples, oranges, strawberries). Eat smaller amounts of *protein* (meat, eggs, milk) and only a very small amount of *fat* (butter, desserts).

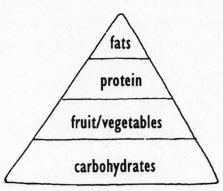

A pyramid is a solid, multisided shape. Most pyramids are built on a three- or four-sided base. Giant pyramids built by early civilizations can still be found today. In Egypt thousands of years ago, pyramid tombs were built to guard the remains and belongings of dead Egyptian kings. In what are now Mexico and Guatemala, an ancient people called the Mayas made pyramid-shaped temples and palaces.

(1) Have your child help you figure out how many total pieces there are of a meal item by multiplying rows by columns (e.g., pieces of chicken or vegetables to be roasted in a pan, or a square pizza to be sliced). Then ask your child such questions as "If we start out with 8 pieces of pizza, and 3 times as many people have to be served, how many pieces are needed?" (2) Your child can organize his or

her toys, clothes, art supplies, etc. in see-through stackable crates. If you arrange them 2 high and 3 wide, how many total crates are there? Show how this is like multiplying rows by columns. If you have your items arranged this way, too (e.g., office supplies, tools, sewing items, dry foods), have your child multiply them the same way.

 My favorite number to multiply 3 by is:

Here's what I like best about multiplying by 3:

4. Let's Go Four-ward

Forest Animals

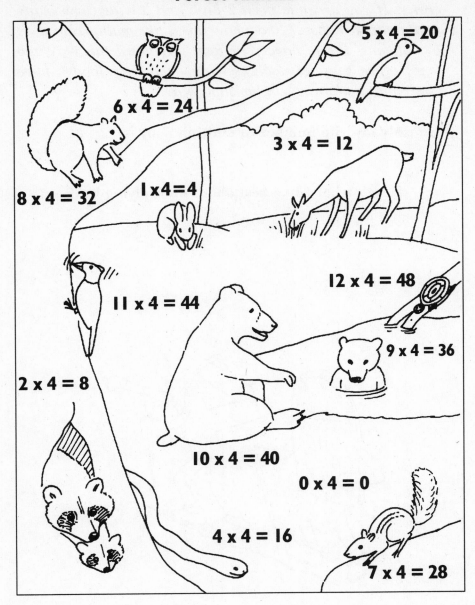

5 x 4 = 20

6 x 4 = 24

3 x 4 = 12

8 x 4 = 32

1 x 4 = 4

12 x 4 = 48

11 x 4 = 44

9 x 4 = 36

2 x 4 = 8

10 x 4 = 40

0 x 4 = 0

4 x 4 = 16

7 x 4 = 28

1. It's easy to multiply any number by 4: The product is just double-double! (Remember that *doubling* means that you add a number to itself.) To multiply by 4, double the number, then double your answer again. For example, to figure out 4 × 3, first double 3 (you get 6), then double your answer to get your product (6 doubled is 12).

3 × 2 = 6
6 × 2 = 12

Four Activity:
Four Corners Cuisine

Cal is visiting the Four Corners, a part of the United States where four states come together. He has discovered a delicious southwestern dish. To find out what it is, do the multiplication exercises in the first column below. If the wrong product is given in the second column, go on to the next exercise. If the correct product is given in the second column, transfer the corresponding letter from the third column into the blanks below, in order.

4 × 1	5	S
4 × 2	8	C
4 × 3	14	P
4 × 4	16	H
4 × 5	25	D
4 × 6	26	F
4 × 7	28	I
4 × 8	30	N
4 × 9	36	L
4 × 10	40	I
4 × 11	42	M
4 × 12	50	Y

_ _ _ _ _

 Can you name the Four Corners states?

 Use multiplication to answer this question as quickly as you can: How many sides are there altogether in this picture?

If your answer was 20, good for you! (There are 4 sides for each square, and 5 squares total. So that's 4 × 5 = 20.)

Become a Mathlete! Exercise with Cal!

Camp Activity

10 x 4 =

4 x 4 =

5 x 4 =

2 x 4 =

12 x 4 =

11 x 4 =

9 x 4 =

3 x 4 =

8 x 4 =

6 x 4 =

1 x 4 =

7 x 4 =

Notes for Myself . . . Just for Fun

Many thousands of years before the first European settlers came to this continent, Native Americans lived here. They belonged to many different tribes, or nations. Some nations in the Southwest included the Hopi, Zuni, Pueblo, and Navajo. Because they lived in the hot, dry desert, they made special houses of a clay called *adobe*. These houses helped them stay cool during the day and warm during the night.

Four is an important number in just about any card game you'll ever play. That's because there are four "suits" in a standard deck of playing cards: clubs, diamonds, hearts, and spades. There are 13 cards in each suit: ace, 2, 3, 4, 5, 6, 7, 8, 9, 10, jack, queen, and king. Altogether, that's a total of 52 cards in each deck (4 suits × 13 cards each = 52 total).

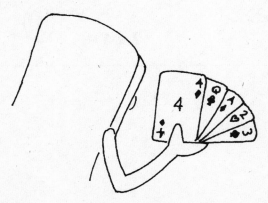

Impress your friends by asking them what 4×13 is. Tell them the answer, then ask if they've ever heard of the game fifty-two pickup. If they shake their heads and say no, here's what you do. Take a full deck of cards, make sure it's all neat and tidy, lift it carefully over your head . . . then flick the cards to the ground as your friends watch in amazement. That's when you say, "Fifty-two cards . . . now pick them up." Then run away—fast!

(1) Many popular board games are designed to be played by up to four players, as reflected by four different playing tokens or by four colors of tokens. Many card games are the same way, either with four individuals or with two teams of two people each. Play these games with your child to encourage cooperation, sportsmanship, strategy, following rules, and, most important, enjoyment. Examples of good beginning board games for four players are Animal Lotto, Parcheesi, Concentration, and dominoes. Examples of good beginning card games for teams are slapjack, old maid, go fish, and snap. (2) Have you ever noticed how the page count for magazines, catalogs, and pamphlets is almost always a multiple of 4? That's because printers work with sheets of paper that are folded four times, then cut to make four separate sheets. In other words, pages are added not one by one, but in groups of four. Next time you and your child are sitting in the waiting room of your dentist's or doctor's office, look through various magazines to see that this is true.

My favorite number to multiply 4 by is:

Here's what I like best about multiplying by 4:

5. Five Does a Dive

Multiplying by 5 is easy. If you multiply an odd number (1, 3, 5, 7, 9, 11) by 5, the product will always end in a 5. If you multiply an even number (2, 4, 6, 8, 10, 12) by 5, the product will always end in 0.

Forest Animals

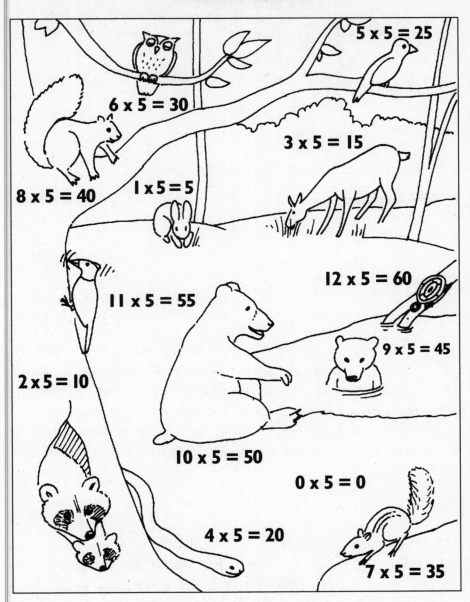

5 x 5 = 25

6 x 5 = 30

3 x 5 = 15

8 x 5 = 40

1 x 5 = 5

12 x 5 = 60

11 x 5 = 55

9 x 5 = 45

2 x 5 = 10

10 x 5 = 50

0 x 5 = 0

4 x 5 = 20

7 x 5 = 35

I'm turning up all over the place!

What pattern do you notice in the following?

Odd Numbers	Even Numbers
1 × 5 = 5	2 × 5 = 10
3 × 5 = 15	4 × 5 = 20
5 × 5 = 25	6 × 5 = 30
7 × 5 = 35	8 × 5 = 40
9 × 5 = 45	10 × 5 = 50
11 × 5 = 55	12 × 5 = 60

An easy way to remember how to multiply a number by 5 is to use the "finger method." Just hold up the same number of fingers as the number you're multiplying by 5, then count those fingers *in groups of 5.* For example, if you're multiplying 6 by 5, hold up 6 fingers, then count in 5s: "Five, ten, fifteen, twenty, twenty-five, thirty" (6 × 5 = 30).

P-s-s-s-t! For another fun way to multiply by 5, skip ahead a few pages to the Magic Act for the number 10.

Camp Activity

10 x 5 =

4 x 5 =

5 x 5 =

2 x 5 =

12 x 5 =

11 x 5 =

9 x 5 =

3 x 5 =

8 x 5 =

6 x 5 =

1 x 5 =

7 x 5 =

Notes for Myself . . . Just for Fun

"I Spy . . . Five." Here's a fun game you can play either indoors or outdoors. Everyone has fifty-five seconds in which to spot anything that there are five of (for example, five picnic tables), or that has five parts (for example, fingers on a hand), then write them down. The person with the most items on his or her list wins.

"Hands Down Winner." It's time to test your skill at skip counting again, and 5s are a good number for doing it with. Play this game with a friend or by yourself, and see how high you can go. Check your progress with this chart:

Skip Counting

If you can count by 5s up to 25	Way to go!
Up to 50	Excellent!
Up to 75	Great job!
Up to 100	Expert skip counter!
Up to 125	You're a true Mathlete!

(1) Anyone who owns a dog or cat has probably been asked, "If your pet is two years old, how old is that in dog (or cat) years?" Help your child learn how to answer this question, first, by explaining that a different means of accounting for time has to be used because the animal's life span is shorter than ours. Then, share with him or her that in general, a dog lives one year for every five of ours. In other words, if a dog is three years old, the child should multiply 3 by 5 to get the dog's relative age of about 15 years. (The commonly held notion that one dog year equals seven human year holds true only for the first couple of years of a dog's life.) If a dog is five years old, how old does that make it in human years? (2) Does your child understand how to take care of a dog, cat, or other pet? Explain to your child that different animals have different needs, such as proper nutrition, good grooming, regular visits to the veterinarian, daily exercise, and lots of love and affection. (3) Wild animals such as squirrels, skunks, etc., do not make good pets. These animals should not be confined to cages. However, it is enjoyable for both the human and the animal if you can set up some sort of feeder for the animal to visit regularly.

 My favorite number to multiply 5 by is:

Here's what I like best about multiplying by 5:

6. Sing Along with Six

Become a Mathlete! Exercise with Callie!

Forest Animals

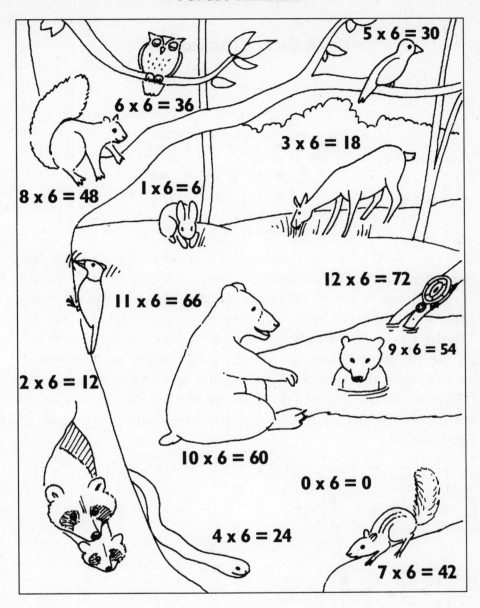

5 x 6 = 30

6 x 6 = 36

3 x 6 = 18

1 x 6 = 6

8 x 6 = 48

12 x 6 = 72

11 x 6 = 66

9 x 6 = 54

2 x 6 = 12

10 x 6 = 60

0 x 6 = 0

4 x 6 = 24

7 x 6 = 42

Six Activity:

A Dicey Situation

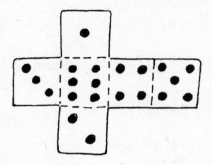

Do you know what dice are? A die is a cube (that means it has six sides). Each side has a certain number of dots on it, from one to six. Some of the world's most popular games are played with one, two, or more dice—can you name any of these games? Maybe you've already played some with your friends or parents!

In the game below, you get to multiply different numbers by 6 and write the product in the blank space. For some of the answers you'll also get to draw the correct number of dots in the blank die or dice. The first one has been done for you as an example.

$6 \times \boxed{\bullet} = \underline{\quad 6 \quad}$

1. $\boxed{\because\cdot} \times \boxed{:::} = \underline{\quad\quad}$

2. $\boxed{:::} \times \boxed{\cdot\,\cdot} = \boxed{} + \boxed{}$

3. ⚃ × ⚅ = ___

4. ⚂ × ⚅ = ▢ + ▢ + ▢

5. 7 × ⚅ = ___

6. ⚅ × 9 = ___

7. 12 × ⚅ = ___

 Here's a shortcut for multiplying even numbers by 6. (Remember that the even numbers we're working with in this book are 2, 4, 6, 8, 10, and 12.)

1. The product will always have two digits.
2. The first digit will always be half of the number you're multiplying by 6.
3. The second digit will always be the number itself.

For example, if you're multiplying 8 by 6, the first digit will be 4 (half of 8), and the second digit will be 8 (the number itself). The product is 48.

8 × 6 = 48

 Multiplying by 6 is easy if you already know how to multiply by 5. Since 6 is 5 + 1, just add the products of multiplying by 5 and 1. For example, let's say you want to multiply 3 by 6.

First, multiply 3 by 5.

3 × 5 = 15

Then multiply 3 by 1.

3 × 1 = 3

Then add the products, and you'll get the same answer as if you had multiplied 3 by 6.

15 + 3 = 18

Here's another example: 9×6.
First, multiply 9 by 5.

9 × 5 = 45

Then multiply 9 by 1.

9 × 1 = 9

Add the products. *Voilà!*

45 + 9 = 54

Become a Mathlete! Exercise with Cal!

Camp Activity

10 x 6 =

4 x 6 =

5 x 6 =

2 x 6 =

12 x 6 =

11 x 6 =

9 x 6 =

3 x 6 =

8 x 6 =

6 x 6 =

1 x 6 =

7 x 6 =

Notes for Myself . . . Just for Fun

More dice tricks for fans of six. Next time you're with one or two of your friends, try this game with a pair of dice. Players take turns rolling the dice. If neither die shows a 6, your score is the sum of the two dice. But if one or both of the dice show a 6, then multiply the numbers for your score. The first player to reach 100 wins.

Multiplication Squares are special puzzles in which all the numbers fit into equations. Once the empty boxes are filled in, each equation can be read correctly either forward, or backward, or up, or down.

Look at the example to see how this works, then fill in the two Multiplication Squares on the next page. As you are filling in the missing numbers, draw arrows to indicate in what direction each equation should read. Note: Some exercises work when you read from top to bottom, others when you read from bottom to top, some when you read from left to right, and others when you read from right to left. But they don't work if you read them diagonally.

Some logic is required to solve these squares; help your children with them if necessary. These squares are a prelude to division.

Squares can be completed in any order that's easy for you. Here's just one example of how you could do it, starting with this Multiplication Square:

6		3
	6	
36	12	

First, you could ask yourself, "Six times what equals 36?" The answer is 6, so write that in.

6		3
6	6	
36	12	

Then you could ask yourself, "Three times what equals 6?" and fill in the square with 2. (Notice that 2 also works reading downward, because 2 × 6 = 2.)

6	2	3
6	6	
36	12	

Then ask yourself, "What times 6 equals 6?" to come up with 1.

6	2	3
6	6	1
36	12	

Finally, ask yourself, "What times 1 equals 3?" to come up with 3. (Notice that 3 also works reading backward across the bottom row, because 3 × 12 = 36.)

6	2	3
6	6	1
36	12	3

And there you have it! Are you ready to do some by yourself now?

2	3	
12		2

		4
2	3	
		24

Five, Six, Pick Up Sticks! "Six" rhymes with lots of other words, like "fix" and "bricks." How many words can you think of that rhyme with "six"? Try to put them all together in a poem, then have a contest with your friends to see who can make the silliest rhymes!

 If you ever find yourself standing in line with your child for a while, such as at a bank or ticket office, play the "Tricks with Six" game. Each person takes a turn to look around and find something, multiply it by 6, and state a sentence that incorporates the product in some way. For example, the first player could spot 2 white boots and say, "I see 12 white boots." The other person then tries to figure out what the first person spotted. (This also prepares your child for division, since he asks himself, "What multiplied by 6 equals 12?")

My favorite number to multiply 6 by is:

Here's what I like best about multiplying by 6:

7. In the Saddle with Seven

Become a Mathlete! Exercise with Callie!

Forest Animals

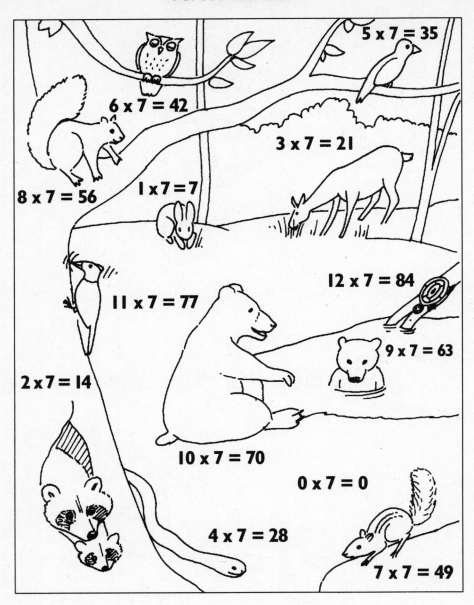

5 x 7 = 35

6 x 7 = 42

3 x 7 = 21

8 x 7 = 56

1 x 7 = 7

12 x 7 = 84

11 x 7 = 77

9 x 7 = 63

2 x 7 = 14

10 x 7 = 70

0 x 7 = 0

4 x 7 = 28

7 x 7 = 49

P-s-s-s-t! Remember that multiplication is fast addition. Skip ahead to some grouping tricks for the number 7.

Seven Activity:

Christmas in the Big Apple

Cal and Callie are so excited! They get to visit Callie's grandmother in New York City for the holidays. Right now they're riding the subway, which is a system of underground trains that take people to different parts of the city. If there are 7 cars in the train, and 2 passengers in each car, how many in all?

7 × 2 = 14

During "rush hour" the subway is more crowded. If the same 7 cars now have 12 passengers each, how many in all?

7 × 12 = 84

Another way of looking at the above exercise is to group it. Remember that 12 = 10 + 2.

7 × 10 = 70
7 × 2 = 14
70 + 14 = 84

It's early in December, and Cal can hardly wait for Christmas. If Christmas is in two weeks, how many more days is that? Use multiplication to get an answer more quickly than addition. (Remember that Christmas is on December 25.)

December						
SUN	MON	TUES	WED	THURS	FRI	SAT
1	2	3	4	5	6	7
8	9	10	11	12	13	14
15	16	17	18	19	20	21
22	23	24	25	26	27	28
29	30	31				

Write your answer in the blanks:

_____ × _____ = _____

☆   If Christmas is three weeks away, what day is it today?

Camp Activity

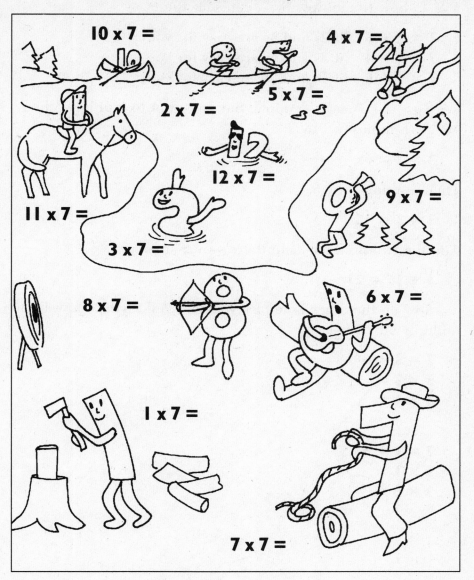

10 x 7 =

4 x 7 =

2 x 7 =

5 x 7 =

12 x 7 =

11 x 7 =

9 x 7 =

3 x 7 =

8 x 7 =

6 x 7 =

1 x 7 =

7 x 7 =

If you want some grouping tricks to help you remember how to multiply by 7, consider these:

7 = 2 + 5   (if 2s and 5s are easy for you)
7 = 3 + 4   (if 3s and 4s are easy for you)
7 = 1 + 6   (if 1s and 6s are easy for you)

Let's use 2 + 5 as an example. Say you want to multiply 3 by 7. First, multiply 3 by 2.

3 × 2 = 6

Then multiply 3 by 5.

3 × 5 = 15

Add the two products, and there's your answer:

6 + 15 = 21

Now do the other examples yourself, still saying you want to multiply 3 by 7.

7 = 3 + 4
3 × 3 = __
3 × 4 = __
__ + __ =

7 = 1 + 6
3 × 1 = __
3 × 6 = __
__ + __ =

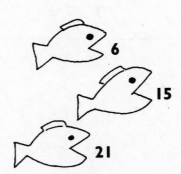

Notes for Myself . . . Just for Fun

Finish this story by choosing words from the list and filling in the blanks. Make it as silly as you want!

Callie's Christmas Present

"Come on, Cal!" said Callie. "Let's go _____ in _____ . It's _____ right now, and the weather should be just perfect!" Cal started to put on his _____ , then stopped. "I thought you wanted to feed _____ _____ at _____ , Callie," he said. Everyone knew these were her favorite _____ because they were so _____ . "Yes, that's true," said Callie, "but we can always go _____ tomorrow, and I know how much you wanted to see _____ because it's _____ ." So they put on their _____ and went _____ at _____ , and it was the best Christmas present they ever had.

mittens
Rockefeller Plaza
7 × 11 floors high
horseback riding
Central Park
skiing
Empire State Building
tourists
ducks
2 × 7 inches small
scuba gear
ice skating
Staten Island
canoeing
7,000 pounds heavy
raining

lions
the Statue of Liberty
snowing
flowers
grumpy
the Bronx Zoo
fishing
things
the Metropolitan Museum of
  Art
cute
8 × 7 feet long
walruses
yodeling
the New York Public Library
animals

If your child is going to spend a week at camp, have him or her figure out how many items of clothing will be needed. For instance, if he or she plans to spend seven nights at camp, then seven shirts will be needed, along with seven pairs of socks, seven T-shirts, etc. How many total items? Then ask your child other questions such as, "If three meals will be eaten per day, how many meals altogether?" or "If you swim twice a day every day, how many times is that?"

My favorite number to multiply 7 by is:

Here's what I like best about multiplying by 7:

8. Don't Be Late, Eight!

Multiplying numbers by 8 is easy. Just think "double-double-double"! In other words, you double (add the number to itself) three times. For example, to figure out 2 × 8, double 2 (you get 4), then double the result (you get 8), then double that result (you get 16).

2 × 2 = 4
4 × 2 = 8
8 × 2 = 16

Here's another example, multiplying 3 by 8. Start by doubling 3, then double the result two more times.

3 × 2 = 6
6 × 2 = 12
12 × 2 = 24

Your answer, 24, is the same as 3 × 8.

To really get the hang of "double-double-double," there's a dice game you can play with two of your friends.

The first person rolls two dice and, whatever the total is, doubles it. The next person doubles that answer. Then the third person doubles it again. Check your answers against the 8 table. Here's an example:

Matthew rolls the dice, gets a total of 5, and doubles it. The answer is 10.

Amelia doubles the 10 to come up with the new answer of 20.

Ian doubles 20 and says the final answer, which is 40.

In other words, 5 × 8 = 40.

Become a Mathlete! Exercise with Cal!

Forest Animals

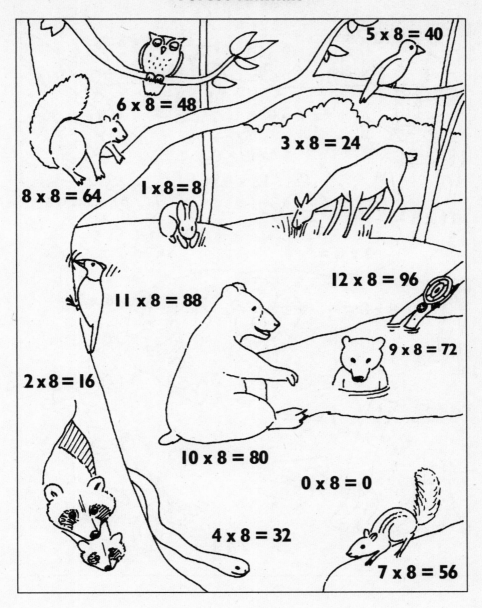

5 x 8 = 40

6 x 8 = 48

3 x 8 = 24

8 x 8 = 64

1 x 8 = 8

11 x 8 = 88

12 x 8 = 96

9 x 8 = 72

2 x 8 = 16

10 x 8 = 80

0 x 8 = 0

4 x 8 = 32

7 x 8 = 56

Camp Activity

10 x 8 =

4 x 8 =

2 x 8 =

5 x 8 =

12 x 8 =

9 x 8 =

11 x 8 =

3 x 8 =

8 x 8 =

6 x 8 =

1 x 8 =

7 x 8 =

Notes for Myself . . . Just for Fun

Eight is an important number in a very ancient game called chess, which came from India fifteen hundred years ago. Chess is played on a game board containing 8 rows and 8 columns. So how many squares does it contain in all? That's right, $8 \times 8 = 64$. Two players, White and Black, play the game with pieces known as pawns, knights, rooks, bishops, queens, and kings. The object of the game is to capture your opponent's king. The pieces move in different ways, with the pawns being least powerful and the queens being most powerful.

(1) Ask your child what is the product of 8 and 10, then share a famous story written by Jules Verne, Around the World in Eighty Days. *Help your child figure out how many weeks and/or months are equal to eighty days. When that book was written, eighty days represented quite an accomplishment! Discuss the various modes of transportation used in the book (e.g., trains, hot-air balloons, steamships) and contrast them with methods used today. How long would it take someone nowadays to circumnavigate the globe? Help your child grasp the concept of time zones and how it was possible for Phileas Fogg to "gain a day" by steadily traveling in the same direction around the globe. Would this still hold true today? (2) Jules Verne was a French author who is considered "the father of science fiction." In his books he predicted the invention of many things, such as space travel and undersea travel. Can your child imagine other things that would make travel even faster in the future?*

 My favorite number to multiply 8 by is:

Here's what I like best about multiplying by 8:

9. Nine is Divine

Nine Activity:

Stargazing

Out in the mountains at Camp Win-with-Math, the sky is so clear
that it's easy to see stars and, sometimes, planets. In fact, Callie
is looking right now at a certain planet. To find out which one
it is, do the multiplication exercises in the first column below,
and circle the correct product from the three choices given. Take
all the letters that correspond to the correct products, and unscramble them.

$9 \times 5 =$	45 (P)	40 (L)	50 (O)
$9 \times 4 =$	39 (S)	36 (R)	34 (T)
$9 \times 8 =$	81 (H)	76 (G)	72 (I)
$9 \times 3 =$	18 (R)	27 (T)	28 (W)
$9 \times 9 =$	90 (L)	81 (J)	82 (I)
$9 \times 6 =$	53 (C)	56 (D)	54 (E)
$9 \times 7 =$	63 (U)	64 (Y)	66 (V)

Forest Animals

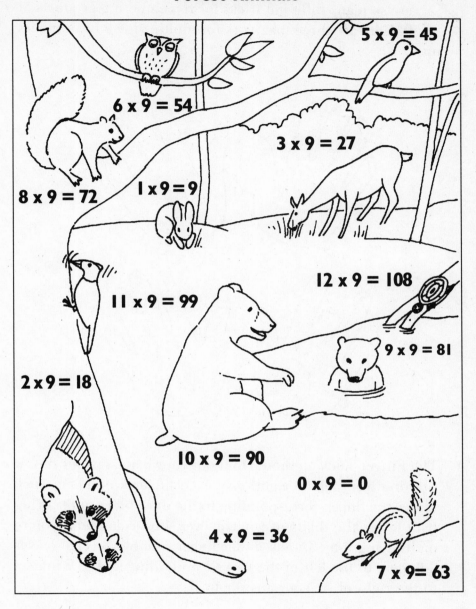

5 x 9 = 45

6 x 9 = 54

3 x 9 = 27

8 x 9 = 72

1 x 9 = 9

12 x 9 = 108

11 x 9 = 99

9 x 9 = 81

2 x 9 = 18

10 x 9 = 90

0 x 9 = 0

4 x 9 = 36

7 x 9 = 63

Nine is the favorite number of lots of kids because there are so many different tricks for remembering products of 9. Here are three that work for multiplying 9 by numbers up to 10:

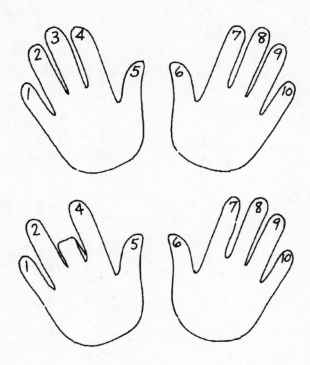

1. The Finger Tuck Method. Hold out your hands, palms down. Starting from left to right, give each finger a number. Tuck under the finger corresponding to the one you're multiplying by 9 to get the digits of your answer. For example, if you're multiplying 9 by 3, tuck under finger number 3. That leaves 2 fingers to the left of the tucked-under finger, and 7 fingers to the right of it. Your answer is 27.

2. Products of 9 always begin with a digit that is one less than the number being multiplied. For example, $9 \times 2 = 18$; $9 \times 3 = 27$.

3. Products of 9 always have digits that add up to 9. Notice the two examples in the previous method.

Camp Activity

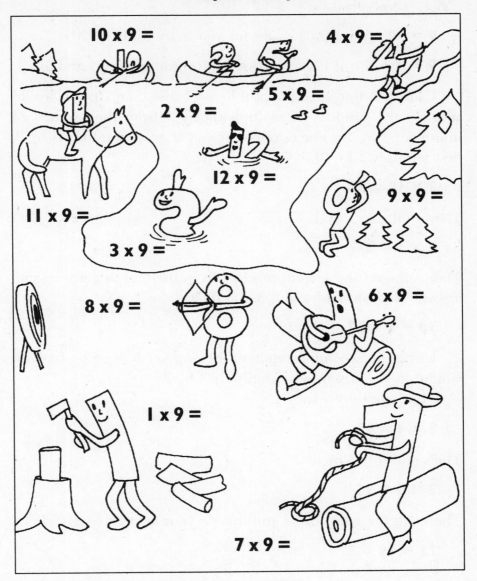

10 x 9 =

4 x 9 =

5 x 9 =

2 x 9 =

12 x 9 =

9 x 9 =

11 x 9 =

3 x 9 =

8 x 9 =

6 x 9 =

1 x 9 =

7 x 9 =

Use grouping to make multiplying by 9 easy. Here are some ideas:

9 = 10 − 1   (if it's easier for you to multiply by 10)

9 = 4 + 5   (if it's easier for you to multiply by 4 and 5)

Using the first example of 9 being 1 less than 10, you would subtract the product of multiplying by 1 from the product of multiplying by 10. For example, if you want to multiply 3 by 9, first multiply 3 by 10.

3 × 10 = 30

Then multiply 3 by 1.

3 × 1 = 3

Then subtract the second product from the first to get the same answer as if you had multiplied 3 by 9.

30 − 3 = 27

Using the second example of 9 being 4 + 5, here's what you would do if you wanted to multiply 3 by 9.

First, multiply 3 by 4.

3 × 4 = 12

Then multiply 3 by 5.

3 × 5 = 15

Then add the two sums—and there's your answer!

12 + 15 = 27

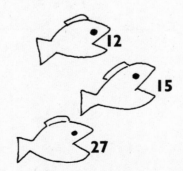

Notes for Myself . . . Just for Fun

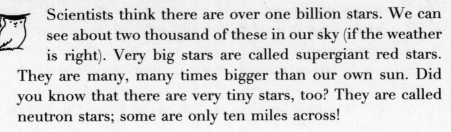 Scientists think there are over one billion stars. We can see about two thousand of these in our sky (if the weather is right). Very big stars are called supergiant red stars. They are many, many times bigger than our own sun. Did you know that there are very tiny stars, too? They are called neutron stars; some are only ten miles across!

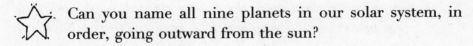 Can you name all nine planets in our solar system, in order, going outward from the sun?

1. Which is the largest?
2. Which is the smallest?
3. Which is known as the "red planet"?
4. Which planets have rings around them?
5. Which two are separated by an asteroid belt?
6. Which planets have moons?

Become a Mathlete! Exercise with Cal!

(1) Show your child where the Big Dipper and other constellations are in the night sky. Stars and their systems are separated by vast distances. After our own sun, the next nearest star to us, Proxima Centauri, is many millions of light-years away. Start imagining with your child: If he or she were the size of a solar system, what would life be like? Would your child use planets as baseballs and stars as flashlights? (2) Now take your imagining in the other direction. If your child were the size of an ant, how big would a spoon be to them? a matchstick? a grain of salt? How about if your child were even smaller? Talk about an atom or about particles that are even smaller than atoms.

My favorite number to multiply 9 by is:

Here's what I like best about multiplying by 9:

10. At–Ten–tion!

Multiplying any number by 10 is easy as can be! Just write the number, then follow it with a 0. For example, if you're multiplying 7 by 10, write 7, then 0. The product is 70. It's 10 times bigger than 7!

Think of Ten as Zero's big brother. They're both very powerful when it comes to multiplication, and very easy to use. Remember that when you multiply any number by 0, the product is always 0. And when you multiply any number by 10, the product always ends with a 0. Pretty impressive!

This is a good time to do some more skip counting. Do it by 10s, either alone or with a friend, and see how high you can go. Then check the chart to see what a whiz you are!

Skip Counting

If you can count by 10s up to 100	You're on your way!
Up to 200	Good going!
Up to 300	Fantastic!
Up to 400	Super sensational!
Up to 500	You're a true Mathlete!

Forest Animals

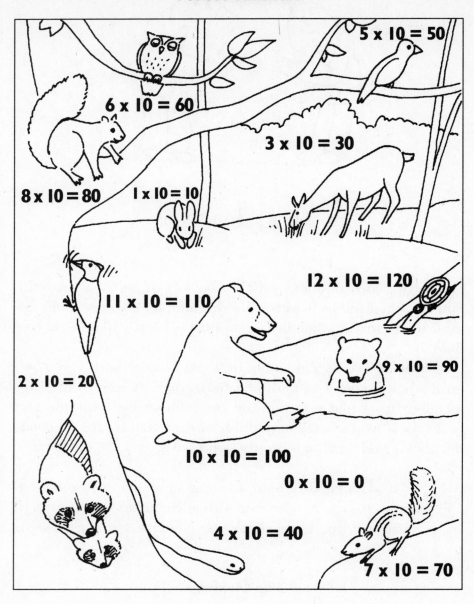

5 x 10 = 50

6 x 10 = 60

3 x 10 = 30

8 x 10 = 80   1 x 10 = 10

12 x 10 = 120

11 x 10 = 110

9 x 10 = 90

2 x 10 = 20

10 x 10 = 100

0 x 10 = 0

4 x 10 = 40

7 x 10 = 70

Camp Activity

10 x 10 =

4 x 10 =

5 x 10 =

2 x 10 =

12 x 10 =

9 x 10 =

11 x 10 =

3 x 10 =

8 x 10 =

6 x 10 =

1 x 10 =

7 x 10 =

Now that you know how to multiply by 10, this gives you another way of multiplying by 5. How? Just remember that, because 5 is half of 10, *any product of 5 is exactly half of the corresponding product of 10.* For example, if $2 \times 10 = 20$, then $2 \times 5 = 10$ (half of 20). Here are some more examples:

$3 \times 10 = 30$, so $3 \times 5 = 15$ (half of 30)
$5 \times 10 = 50$, so $5 \times 5 = 25$ (half of 50)
$9 \times 10 = 90$, so $9 \times 5 = 45$ (half of 90)
$10 \times 10 = 100$, so $10 \times 5 = 50$ (half of 100)

Check this out by doing your skip counting again. Write down all the numbers you get when you count by 5s, then do the same thing for 10s, then compare them. See the pattern?

Notes for Myself . . . Just for Fun

Uh-oh! Ten is supposed to make sure that all the sports equipment has been returned to the shed, but notices that 2×10 items are missing. How many is that? Also, can you spot the hidden objects? Find: 1 baseball bat, 2 baseballs, 1 snorkel, 1 dive mask, 2 fins, 1 bow, 5 arrows, 1 pair of oars, 1 soccer ball, 1 Indian headdress, 2 moccasins, 1 pair of binoculars, and 1 magnifying glass.

Become a Mathlete! Exercise with Cal!

 Whether you work at home, bring your work home on the weekends, or occasionally take your child with you to work, your child may be able to help you. Younger children can sort items such as paperclips, pens, rubberbands, note paper, etc., into neat groups of 10 each, then tell you the total number of items. Older children who are more responsible can make photocopies, figuring out beforehand how many total pages there will be, then collate. For example, if you needed 10 collated copies of a 4-page document, how many total pages would there be? They can also

help you stuff envelopes, or fold newsletters or brochures neatly, and put them in groups of 10 to facilitate getting a final count before mailing.

 My favorite number to multiply 10 by is:

Here's what I like best about multiplying by 10:

11. Eagle Eye Eleven

On the opposite page, what pattern do you notice for products of 11? That's right . . . for the numbers 1 through 9, the product is simply that digit written twice! For example, 2 × 11 = 22, 3 × 11 = 33, 4 × 11 = 44, and so on. Easy as pie! (This works for all single-digit numbers: 1, 2, 3, 4, 5, 6, 7, 8, and 9.)

Forest Animals

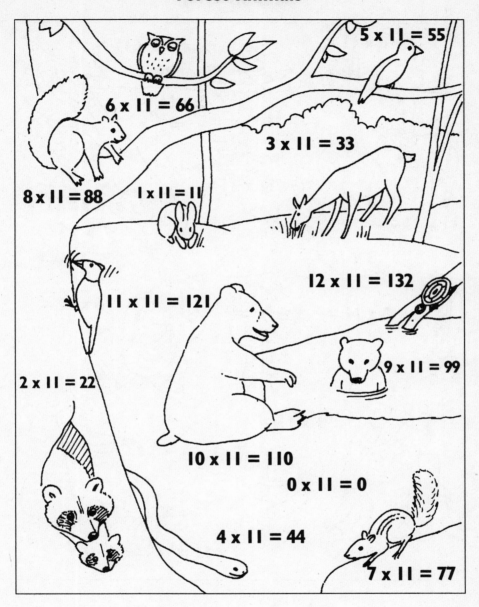

5 x 11 = 55

6 x 11 = 66

3 x 11 = 33

8 x 11 = 88

1 x 11 = 11

12 x 11 = 132

11 x 11 = 121

9 x 11 = 99

2 x 11 = 22

10 x 11 = 110

0 x 11 = 0

4 x 11 = 44

7 x 11 = 77

Camp Activity

10 x 11 =

4 x 11 =

5 x 11 =

2 x 11 =

12 x 11 =

11 x 11 =

9 x 11 =

3 x 11 =

8 x 11 =

6 x 11 =

1 x 11 =

7 x 11 =

Here's a way to remember the products of 11 multiplied by two-digit numbers such as 10, 11, or 12. Just space the digits apart, and in between them write their sum — and that's the answer! For example:

10 × 11

Space the two digits of 10 apart, and write them down:
   1 __ 0

Add the two digits. 1 + 0 = 1, so write this sum in between the digits: 1 **1** 0

10 × 11 = 110

Let's do it again with 11.

11 × 11

Space the two digits of 11 apart: 1 __ 1

1 + 1 = 2. Write this sum in between the 2 digits: 1 **2** 1

11 × 11 = 121

Now do it yourself with 12.

12 × 11

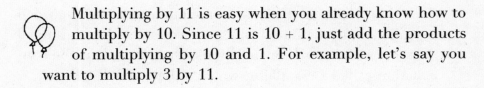

___ ___ ___

___ + ___ = ___

12 × 11 = _____

Multiplying by 11 is easy when you already know how to multiply by 10. Since 11 is 10 + 1, just add the products of multiplying by 10 and 1. For example, let's say you want to multiply 3 by 11.

First, multiply 3 by 10.

3 × 10 = 30

Then multiply 3 by 1.

3 × 1 = 3

Then add the products, and you'll get the same answer as if you had multiplied 3 by 11.

30 + 3 = 33

What a cinch!

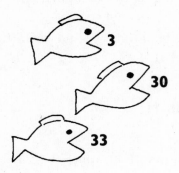

Notes for Myself . . . Just for Fun

Eleven multiplied by 8 gives the product of 88. This number is important for at least a couple of reasons. One has to do with pianos. There are 88 keys, or 11 octaves (groups of 8 notes) on a full-sized piano keyboard—that's a lot of notes!

The number 88 is also important to astronomers—scientists who study our solar system and other solar systems. From the beginning of our history, people have watched the sky and noticed that the stars seemed to form pictures, which are called constellations. Modern astronomers have grouped stars into 88 of these constellations. One example is the constellation called Ursa Major, which is Latin for Big Bear.

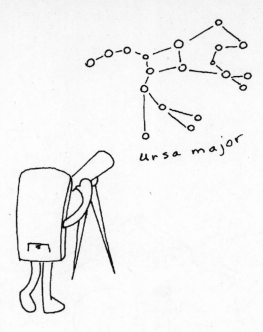

ursa major

 Here's a game you can play with your child the next time you are out together somewhere, such as at a shopping mall or a park. Say, for example, that you're at a pet store. Pick an object and say, "I see a mouse. If it were 11 times bigger, what would it be?" and your child could answer, "An elephant." Then he or she gets a turn. When it's your turn again, say, "I see 3 rawhide bones. If there were 11 times more, how many would there be?" and your child can give you the total. You can also ask what would be 11 times smaller, faster, slower, louder, brighter, and so on. For younger children, concentrating on the questions having to do with quantities is best, especially if you can manipulate objects such as beads, paper clips, etc.

My favorite number to multiply 11 by is:

Here's what I like best about multiplying by 11:

12. Tap-dancing Twelve

Forest Animals

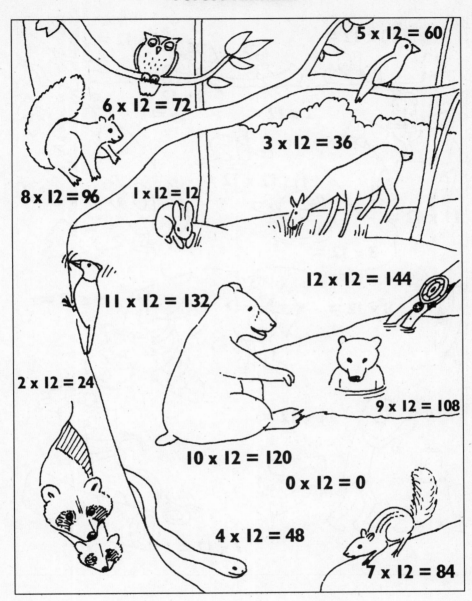

5 x 12 = 60

6 x 12 = 72

3 x 12 = 36

8 x 12 = 96

1 x 12 = 12

12 x 12 = 144

11 x 12 = 132

2 x 12 = 24

9 x 12 = 108

10 x 12 = 120

0 x 12 = 0

4 x 12 = 48

7 x 12 = 84

Camp Activity

10 x 12 =

4 x 12 =

5 x 12 =

2 x 12 =

12 x 12 =

9 x 12 =

11 x 12 =

3 x 12 =

8 x 12 =

6 x 12 =

1 x 12 =

7 x 12 =

Encourage your children to find patterns wherever they can in the multiplication tables, such as the one for 12. You could point out to them, for example, how the last digit of each product grows steadily by 2s (i.e., 2, 4, 6, 8, 10, 12 . . .). This leads eventually to a discussion of grouping as another way to learn the 12s. For example: $1 \times 12 = (1 \times 10) + (1 \times 2)$, $2 \times 12 = (2 \times 10) + (2 \times 2)$, and so on. . . . Look for the grouping icon below.

Did you notice that some of the products of 12 are numbers we use in keeping track of time? Answer these questions:

1. An old-fashioned clock has ___ numbers on it, or ___ × 12.

2. There are ___ hours in a day, or ___ × 12.

3. There are ___ seconds in a minute and ___ minutes in an hour, or ___ × 12.

 Since you already know how to multiply by 2, and also how to multiply by 10, multiplying by 12 is easy. Just remember that 12 is the same as 10 + 2, so you can get the same result by adding the products of multiplying by 10 and 2. For example, let's say you want to multiply 3 by 12.

First, multiply 3 by 10.

3 × 10 = 30

Then multiply 3 by 2.

3 × 2 = 6

Then add the products, and you'll get the same answer as if you had multiplied 3 by 12.

30 + 6 = 36

Way too easy!

Now try it with 4 in order to find the product of 4 × 12.

4 × 10 = __
4 × 2 = __
__ + __ = __

Notes for Myself . . . Just for Fun

It's Talent Night at Camp Win-with-Math, and the Amazing Cal is astonishing everyone with his mind-reading act. Right now he's doing a card trick. Here's how you can amaze your friends, too:

1. Remove all the 10s and face cards from a deck of cards.

2. Have a friend shuffle the deck, then choose two cards.

3. After looking at both cards, your friend gives you one of them. Look at your card, but not your friend's.

4. Have your friend multiply the number on his or her card by 2 (aces = 1), then add 2 to the result, then multiply the newest result by 5.

5. From the chart below, tell your friend the Special Number for your card, and have him or her subtract the Special Number from his or her answer.

6. Take back your friend's card and hold it, face showing, to the left of yours. The two cards should be the digits of your friend's final answer!*

*Don Fraser, "A Card Trick," in *Mathemagic* (Boston: Addison-Wesley Publishers, 1985), p. 42.

For example:

1. Let's say your friend has a 6, and you have a 2.

2. Your friend multiples 6 by 2 (to get 12), adds 2 to 12 (to get 14), then multiplies 14 by 5 (to get 70). (Use grouping to do this more easily: $14 = 10 + 4$; $10 \times 5 = 50$; $4 \times 5 = 20$; $50 + 20 = 70$.)

3. The Special Number for your card, 2, is 8. Your friend subtracts 8 from his or her number, 70, to get 62 — which matches your cards, 6 and 2!

Your Card	Special No.
1	9
2	8
3	7
4	6
5	5
6	4
7	3
8	2
9	1

 Does your child know the Christmas carol "The Twelve Days of Christmas"? ("On the first day of Christmas my true love gave to me a partridge in a pear tree/On the second day of Christmas my true love gave to me two turtledoves and a partridge in a pear tree . . .") Younger children enjoy the repetition of the numbers. Older children can help you figure out how quickly all the gift items add up with each successive day. At the end of the twelve days, how many total items will there be? Urge your child to use multiplication to help you solve this problem ([12 × 1 partridge] + [11 × 2 turtledoves] + [10 × 3 French hens] . . .).

My favorite number to multiply 12 by is:

Here's what I like best about multiplying by 12:

Multiplication Review

Congratulations! You have learned a lot about multiplication. The first thing you learned is that multiplication is really *fast addition*. Then you learned that the answer is called a *product*. In multiplication, just as in addition, it doesn't matter in what order you multiply the numbers because the answer is always the same.

Numbers do all kinds of interesting things when they're multiplied. Zero is a powerful number in multiplication because when you multiply any number by 0, the product is 0. If you multiply a number by 1, the number stays the same. Any number multiplied by 2 doubles, which is just like adding the number to itself. When you multiply a number by 4, it's "double-double," and when you multiply a number by 8, it's "double-double-double." To multiply a number by 10, all you do is write a 0 after the number.

In this chapter you started a great new game called skip counting, where you counted by 2s, 3s, 5s, and 10s. Skip counting is a super shortcut! How high were you able to go? Did you race your friends?

Another multiplication trick you learned was the "row/column" method. Let's say you have a pan of brownies, and you want to know exactly how many there are. Simply count how many pieces there are in a row (let's say there are 3), then how many pieces there are in a column (let's say 5). Multiply 3 by 5, and that's how many brownies are in the pan: 15.

Grouping came in handy for multiplication, just as it did for addition. Grouping lets you use what you already know about certain numbers to make it easier to multiply by other numbers. For example, if you're not sure about multiplying by 12, you can "rearrange" 12 into 2 + 10. Multiply your number by 2, then by 10, and add the sums—and the product is the same as if you had multiplied by 12 to begin with!

Multiplication Activity: It's Time for a Times Table

How well do you know your multiplication? Fill in each of these squares with the correct product, multiplying each number in the left column by each number in the top row, until all the squares have been filled in. Have someone time you to see how fast you did it. Then check your answers against the Multiplication Table on page 247. Good luck!

Multiplication Table

×	0	1	2	3	4	5	6	7	8	9	10	11	12
0	0												
1		1											
2		2	4										
3													
4													
5													
6													
7													
8													
9													
10													
11													
12													

Following the examples, do each of the pairs of multiplication exercises below to find out which is greater. If the two are the same, write an equals sign between them.

Examples:

7 × 3 < 2 × 11   (21 is less than 22)

2 × 1 > 0 × 12   (2 is greater than 0)

4 × 6 = 12 × 2   (24 equals 24)

1.	4 × 10	8 × 5
2.	5 × 3	2 × 7
3.	9 × 6	7 × 8
4.	3 × 11	5 × sides on a cube
5.	6 × 8	7 × 7
6.	2 × 9	3 × 6
7.	10 × 10	12 × 12
8.	3 × 3	2 × 4
9.	eggs in a dozen	7 × 2
10.	9 × 3	5 × 5
11.	12 × 9	11 × 10
12.	5 × 12	10 × 6
13.	legs on a horse × 11	6 × days in a week
14.	6 × 6	3 × 12
15.	1 × 1	0 × 2
16.	arms on an octopus × 8	9 × 7
17.	3 × 9	4 × 7
18.	11 × 11	12 × 10
19.	8 × 3	6 × 4
20.	12 × 6	9 × 8

As you read across the rows, do you notice anything familiar? If you've been doing your skip counting, then you should recognize several of these rows, such as the 2s, 3s, 5s, and 10s. This Multiplication Table gives you a good start for skip counting other numbers as well—try it and see!

Multiplication Table

×	0	1	2	3	4	5	6	7	8	9	10	11	12
0	0	0	0	0	0	0	0	0	0	0	0	0	0
1	0	1	2	3	4	5	6	7	8	9	10	11	12
2	0	2	4	6	8	10	12	14	16	18	20	22	24
3	0	3	6	9	12	15	18	21	24	27	30	33	36
4	0	4	8	12	16	20	24	28	32	36	40	44	48
5	0	5	10	15	20	25	30	35	40	45	50	55	60
6	0	6	12	18	24	30	36	42	48	54	60	66	72
7	0	7	14	21	28	35	42	49	56	63	70	77	84
8	0	8	16	24	32	40	48	56	64	72	80	88	96
9	0	9	18	27	36	45	54	63	72	81	90	99	118
10	0	10	20	30	40	50	60	70	80	90	100	110	120
11	0	11	22	33	44	55	66	77	88	99	110	121	132
12	0	12	24	36	48	60	72	84	96	108	120	132	144

6

Share and Share Alike

Division with the Human Calculator

Division is *reverse multiplication.* Since you already know multiplication, you know a lot about division! They are like two sides of the same coin. Saying "Three multiplied by what equals 6?" ($3 \times ? = 6$) is the same as saying "Six divided by 3 equals what?" ($6 \div 3 = ?$).

In division, the answer is called a *quotient.* For example, in the division equation $8 \div 2 = 4$, the *quotient* is 4. Circle the quotients in the following:

$$12 \div 6 = 2$$
$$7 \div 7 = 1$$
$$10 \div 2 = 5$$
$$8 \div 1 = 8$$

That's right! The quotients are 2, 1, 5, and 8.

Division is nifty. It lets you split things up — *divide* them — into equal groups or parts. For example, let's say that Dave, Carla, and Ariana are eating lunch. There are 3 cartons of juice on the table, and they want to share equally. That's an easy one to figure out: 3 (juice cartons) divided by 3 (people) equals 1 (juice carton each). Or, $3 \div 3 = 1$. Each person gets exactly 1 juice carton.

By the way, you'll notice that 0 does not show up in this chapter. That's because you can't divide by 0. Why not? Because you can't say, for example, "Zero multiplied by what equals 3?" since this is never true.

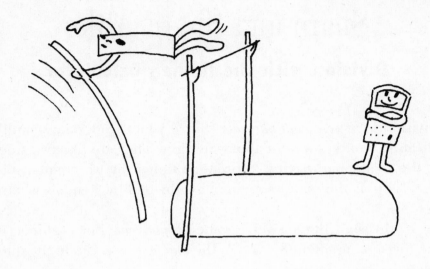

1. One's Solo Act

Just like multiplication (where any number multiplied by 1 stays the same), any number *divided* by 1 stays the same. For example:

$$1 \div 1 = 1$$
$$2 \div 1 = 2$$
$$3 \div 1 = 3$$
$$10 \div 1 = 10$$
$$xyz \div 1 = xyz$$
$$4{,}567{,}843{,}000 \div 1 = 4{,}567{,}843{,}000$$

Reptile/Dinosaur/Amphibian

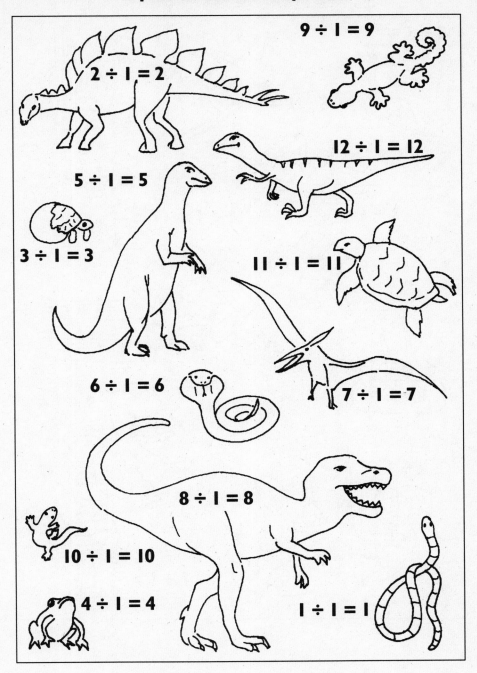

$9 \div 1 = 9$

$2 \div 1 = 2$

$12 \div 1 = 12$

$5 \div 1 = 5$

$3 \div 1 = 3$

$11 \div 1 = 11$

$6 \div 1 = 6$

$7 \div 1 = 7$

$8 \div 1 = 8$

$10 \div 1 = 10$

$4 \div 1 = 4$

$1 \div 1 = 1$

Individual Sports

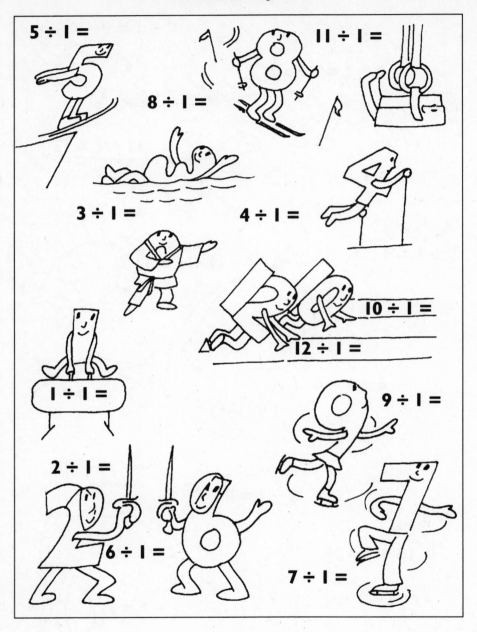

5 ÷ 1 =

8 ÷ 1 =

11 ÷ 1 =

3 ÷ 1 =

4 ÷ 1 =

10 ÷ 1 =

12 ÷ 1 =

1 ÷ 1 =

9 ÷ 1 =

2 ÷ 1 =

6 ÷ 1 =

7 ÷ 1 =

Remember, any number divided by 1 stays the same. And here's something else to know: If you divide any number *by itself* the quotient will always be 1. For example:

$1 \div 1 = 1$
$2 \div 2 = 1$
$3 \div 3 = 1$
$10 \div 10 = 1$
$xyz \div xyz = 1$
$4{,}567{,}843{,}000 \div 4{,}567{,}843{,}000 = 1$

Solve the following:

Become a Mathlete! Exercise with Cal!

Notes for Myself . . . Just for Fun

Mexico is south of the United States and has 83 million people living there. The capital, Mexico City, is the world's most populated city, with 15 million people. Oil and minerals are important resources, and many people make their living by working in factories or farming such crops as cotton, sugar, coffee, and tomatoes. The unit of currency is the Mexican peso.*

The official language of Mexico is Spanish. The word for "one" in Spanish is *uno* (pronounced OOOOH-no).

*"Mexico," *The Random House Children's Encyclopedia* (New York: Random House, 1991), pp. 340–341.

Discuss the trick-question word problem above with your child to show that actually there is an answer. (Brandon, Jessica, and Stephen would each get an equal number of candy pieces.)

Many sports activities build individual skills and are rewarding to do by oneself. They can be done in competitive environments or simply for enjoyment. These include bicycling, surfing, gymnastics, skateboarding, running, track and field events, and many others. Get in the habit of spending time walking, jogging, or bicycling with your child, as well as encouraging him or her to participate in these activities.

My favorite number to divide by 1 is:

I like dividing by 1 because:

2. Easy Two Do

Remember that division is reverse multiplication. When you multiply a number by 2, you double it. But when you *divide* a number by 2, you *halve* it. For example, if you are dividing 10 pencils up between 2 people (that is, dividing 10 by 2), each person gets *half* the original amount: 5.

Become a Mathlete! Exercise with Callie!

Reptile/Dinosaur/Amphibian

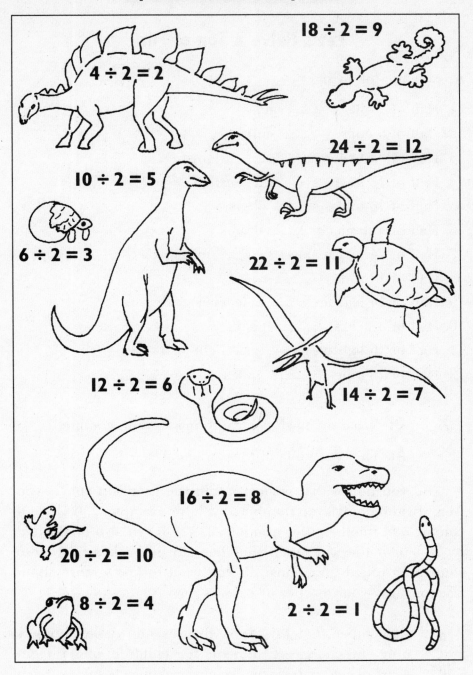

$18 \div 2 = 9$

$4 \div 2 = 2$

$24 \div 2 = 12$

$10 \div 2 = 5$

$6 \div 2 = 3$

$22 \div 2 = 11$

$12 \div 2 = 6$

$14 \div 2 = 7$

$16 \div 2 = 8$

$20 \div 2 = 10$

$8 \div 2 = 4$

$2 \div 2 = 1$

Two Activity:

Let's Halve a Ton of Fun

Answer the following:

1. Half of 2 gills = _____ gill
2. Half of 6 cups = _____ cups
3. Half of 22 furlongs = _____ furlongs
4. Half of 18 bushels = _____ bushels
5. Half of 10 liters = _____ liters
6. Half of 4 drams = _____ drams
7. Half of 24 megaparsecs = _____ megaparsecs
8. Half of 12 rods = _____ rods
9. Half of 14 leagues = _____ leagues
10. Half of 8 pecks = _____ pecks
11. Half of 16 light-years = _____ light-years
12. Half of 20 miles = _____ miles

Q: What do all the above things have in common?

A: They are units of measurement

Did you notice that when you did the exercises above, you were dividing only even numbers by 2 (for example, 2, 6, 22)? Of course, odd numbers (for example, 1, 5, 13) can also be divided by 2—but if they are, then they have to be broken apart into something called "fractions." If you want to know more about fractions, ask your teacher or parents.

Remember that this book focuses on whole numbers only. Fractions are difficult for children under eight years of age, unless you use physical objects such as measuring cups to illustrate the concept.

Individual Sports

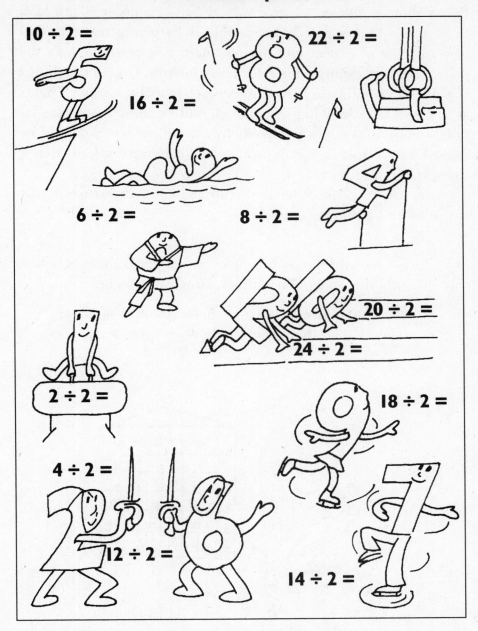

10 ÷ 2 =

22 ÷ 2 =

16 ÷ 2 =

6 ÷ 2 =

8 ÷ 2 =

20 ÷ 2 =

24 ÷ 2 =

2 ÷ 2 =

18 ÷ 2 =

4 ÷ 2 =

12 ÷ 2 =

14 ÷ 2 =

Notes for Myself . . . Just for Fun

Being able to *estimate,* or give a good guess, is an excellent skill to practice. One way it helps people is by giving them a good idea of what to expect and how to plan. For example, let's say your mom is planning a birthday party for you. One of the things she'll have to estimate is how many guests will show up. This is so she can plan how big the cake should be, how many bags of potato chips and other goodies to buy, whether games should be played inside the house or out in the backyard where there's more room, and so on.

Can you think of other things to estimate, and reasons why it would be helpful?

Q: Samantha was born eleven years ago, but she's had only three actual birthdays. How can this be?

A: Samantha was born on February 29 — in a leap year — a date that occurs only once every four years.

(1) After a shopping trip, have your child help you put groceries away. For example, if you have 6 large grapefruits and 2 small bowls to put them in, how many grapefruits can the child put in each bowl? Or, if you are putting laundry away and there are 18 T-shirts to be divided up between 2 drawers, how many go in each drawer? (2) Ask your child how many children are in his or her class, then to halve that number (if there is an odd number of children, include the teacher in the calculation). Or if there are 12 children at a party, and they were to form 2 lines, how many would be in each line?

My favorite number to divide by 2 is:

I like dividing by 2 because:

3. Three Cheers for Three!

Now you're an expert at dividing by 1 and 2! Easy, aren't they? And when the numbers start getting larger, division can *really* come in handy. For example, let's say you're eating lunch with 2 of your friends. You have 15 pretzel sticks you want to share equally among you. Remember that division is reverse multiplication. You can either ask yourself, "Fifteen (pretzel sticks) divided by 3 (people) equals what?" or "Three (people) times what equals 15 (pretzel sticks)?"

$$15 \div 3 = 5$$
$$3 \times 5 = 15$$

Either way, the answer is 5. Each of you will get 5 pretzel sticks. Aren't you glad you know how to divide?

Reptile/Dinosaur/Amphibian

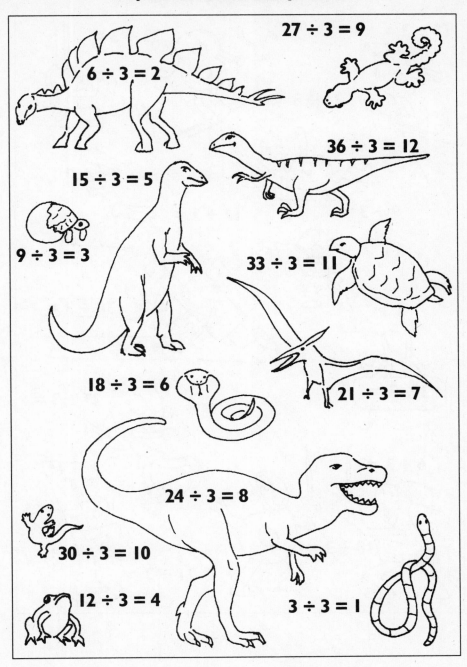

$27 \div 3 = 9$

$6 \div 3 = 2$

$15 \div 3 = 5$

$36 \div 3 = 12$

$9 \div 3 = 3$

$33 \div 3 = 11$

$18 \div 3 = 6$

$21 \div 3 = 7$

$24 \div 3 = 8$

$30 \div 3 = 10$

$12 \div 3 = 4$

$3 \div 3 = 1$

Individual Sports

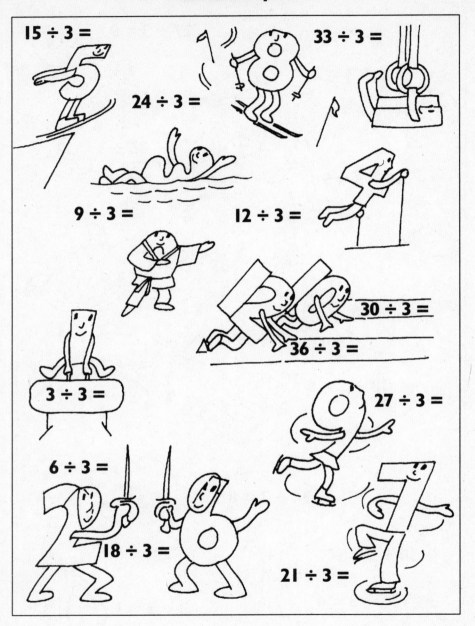

15 ÷ 3 =

24 ÷ 3 =

33 ÷ 3 =

9 ÷ 3 =

12 ÷ 3 =

30 ÷ 3 =

36 ÷ 3 =

3 ÷ 3 =

27 ÷ 3 =

6 ÷ 3 =

18 ÷ 3 =

21 ÷ 3 =

Notes for Myself . . . Just for Fun

 Do you use a computer at home or at school? Computers help us to learn in fun new ways. They also let us play games—which ones are your favorites?

Early types of computers, or calculating machines, have been used for thousands of years. For example, the abacus, made of a frame and rows of beads, was invented in China and is still used today. In the seventeenth century, a simpler type of computer called a calculator was used. Today's computers are special kinds of machines that work a little bit like our own brains. Although there isn't yet a computer as smart as the human brain, scientists are working on it so that computers can help us even more than they already do.

A computer has memory for storing information and programs (instructions). It also has a CPU (central processing unit) that lets it do such things as dividing numbers rapidly or creating words and pictures. For example, you could give the computer some numbers to divide (input); then its CPU will do the division problem and give you the answer (output).*

*"Computers," *The Random House Children's Encyclopedia* (New York: Random House, 1991), pp. 135–136.

You can think of your brain as a supersmart computer. You're always getting "input," whenever you see, hear, touch, taste, or smell something. The things you think, do, and say are the "output."

☆ What is the correct output (quotient) in these exercises?

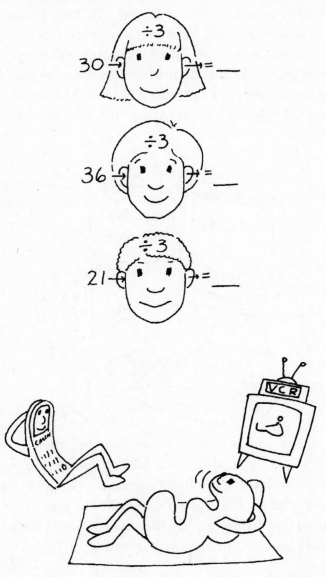

$$30 \div 3 = \underline{\quad}$$

$$36 \div 3 = \underline{\quad}$$

$$21 \div 3 = \underline{\quad}$$

Become a Mathlete! Exercise with Callie!

Share with your child that different countries use different currencies than we do, such as the peso in Mexico and the yen in Japan, and that these currencies do not correspond one-on-one with ours. Pretend with your child that you are in Mexico and that it takes 3 Mexican pesos to equal 1 American dollar. If you see a wallet that is selling for 15 pesos, how much is that in dollars? Your child would divide 15 by 3 to give you the answer. Do this with several other examples. For older children, you may wish to discuss how some currencies are "stronger" than others by having more buying power. If the situation above were reversed, and it took 3 American dollars to equal 1 of a different type of currency, would the dollar be considered stronger or weaker?

My favorite number to divide by 3 is:

I like dividing by 3 because:

4. Fencing Four

We're already up to 4! Do you remember what division is? That's right—reverse multiplication. If you're dividing 48 by 4, you can think of it as reverse multiplication: "Four multiplied by what equals 48?"

$$48 \div 4 = 12$$
$$4 \times 12 = 48$$

Either way, the answer is 12.

A shortcut for dividing by 4 is "half-half." Just take half of the number you're dividing, and halve it again. For example, let's say you're dividing 12 by 4. Half of 12 is 6, and half of 6 is 3—so $12 \div 4 = 3$.

Do it yourself with $20 \div 4$.

Half of 20 is _____ .
Half of _____ is _____ .
$20 \div 4 =$ _____ .

Reptile/Dinosaur/Amphibian

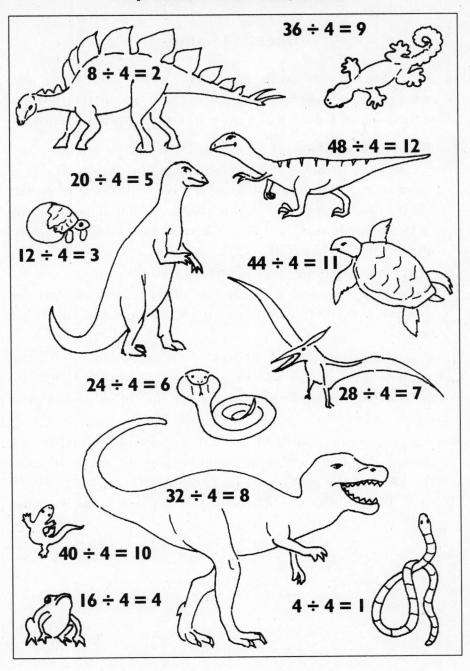

$36 \div 4 = 9$

$8 \div 4 = 2$

$48 \div 4 = 12$

$20 \div 4 = 5$

$12 \div 4 = 3$

$44 \div 4 = 11$

$24 \div 4 = 6$

$28 \div 4 = 7$

$32 \div 4 = 8$

$40 \div 4 = 10$

$16 \div 4 = 4$

$4 \div 4 = 1$

Four Activity:

Guessing Game

Here is a game to play with your friends. Time yourselves, and see who can come up with the correct answers first. After you solve the puzzles below, make up your own for other numbers.

1. This is the quotient of 44 and 4. It's a 2-digit number, and it's also the same as the quotient of 8 and 4, plus 9. It's 4 more than the number of days in a week. It's one less than the number of eggs in a carton. (Remember that another way of thinking about 44 ÷ 4 is to ask yourself, "Four times what equals 44?") What is it?

2. This is the quotient of 24 and 4. It's an even number. It's the same as 2 × 3, and it's only one digit. It rhymes with the material that the third little pig made his house out of. What is it?

3. This is the quotient of 36 and 4. An odd number, it's the same as the number of toes on your feet, minus 1. It's the same as 16 divided by 4, plus the number of sides in a pentagon. What is it?

4. This is the quotient of 32 and 4. It's a one-digit number. It's the same as the number of piano keys in an octave. It's also the same as the number of legs on a spider. If you divide 16 by 4, then double the quotient, you get this number. What is it?

Individual Sports

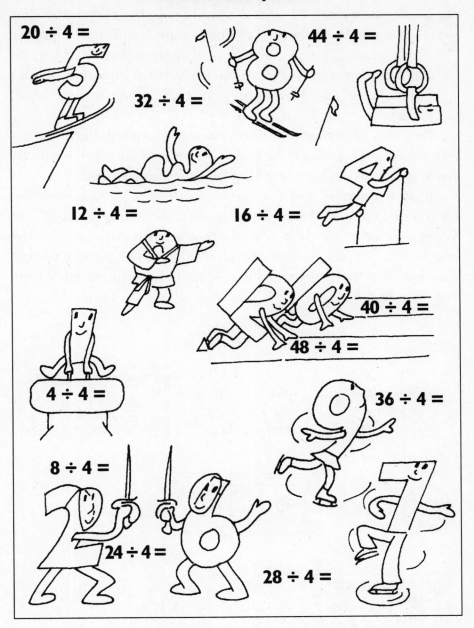

$20 \div 4 =$

$44 \div 4 =$

$32 \div 4 =$

$12 \div 4 =$

$16 \div 4 =$

$40 \div 4 =$

$48 \div 4 =$

$4 \div 4 =$

$36 \div 4 =$

$8 \div 4 =$

$24 \div 4 =$

$28 \div 4 =$

Notes for Myself . . . Just for Fun

It takes four of a certain kind of coin to make one dollar. Do you know what it is? That's right, a quarter! A quarter is worth 25 cents, and four of them together add up to 100 cents, or one dollar. (In fact, the word "quarter" comes from a Latin word meaning "fourth," or one of four equal parts.)

One way of saying that four quarters make a dollar is with a multiplication equation: $25 \times 4 = 100$. Can you reverse this by creating a division equation that says the same thing?

Find a quarter, and compare it with other coins: pennies, nickels, and dimes. Which is the biggest? Which is the smallest? What type of metal do you think they're made of? Look at the edges of the coins. Which are grooved, and why do you think they are? Are there any coins that seem to be made of more than one type of metal?

(1) Has your child ever heard the joke about Silly Sam? He was visiting a cattle ranch one day, and the rancher asked him, "Do you know how many cattle I have?" and without a moment's pause, Silly Sam said, "Forty-eight." The rancher said, "That's amazing! How did you count them up so fast?" Silly Sam said, "Simple. I counted all the legs, then divided by four!" See if your child can be another Silly Sam. Take turns with one person looking through picture books or magazines and finding a four-legged animal (or something else with four parts), then saying how many legs he or she sees and asking the other person the correct number of animals. (2) Do other conversions to four with your child. For example, if there are 4 seasons in 1 year, and there are 16 seasons, how many years is that? Or, if there are 4 quarts in a gallon, if you have 8 quarts, how many gallons is that? Or, if 4 quarters equal 1 dollar, if you have 24 quarters, how many dollars do you have?

My favorite number to divide by 4 is:

I like dividing by 4 because:

Become a Mathlete! Exercise with Cal!

5. High Five

Become a Mathlete! Exercise with Callie!

Reptile/Dinosaur/Amphibian

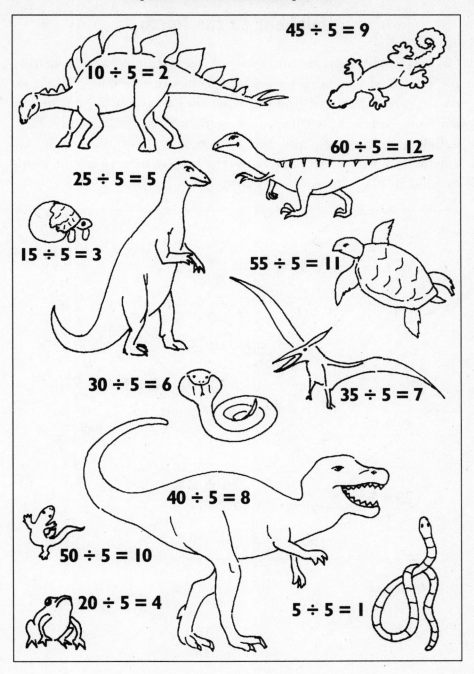

$45 \div 5 = 9$

$10 \div 5 = 2$

$60 \div 5 = 12$

$25 \div 5 = 5$

$15 \div 5 = 3$

$55 \div 5 = 11$

$30 \div 5 = 6$

$35 \div 5 = 7$

$40 \div 5 = 8$

$50 \div 5 = 10$

$20 \div 5 = 4$

$5 \div 5 = 1$

Five Activity:

Neighbor to the North

Cal can hardly wait! He and some of his Mathlete pals are visiting the city of Montreal in Canada, and they're eager to play Canada's favorite sport. To find out what they're going to play, do the division exercises in the spaces below. Then look at the chart, which matches each quotient to a color. Color the spaces according to this chart, and you'll see a picture of some items needed to play this mystery sport.

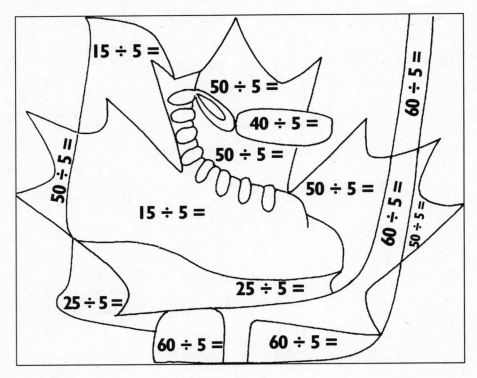

Quotient	Color
10	red
3	black
12	brown
5	gray
8	black

Individual Sports

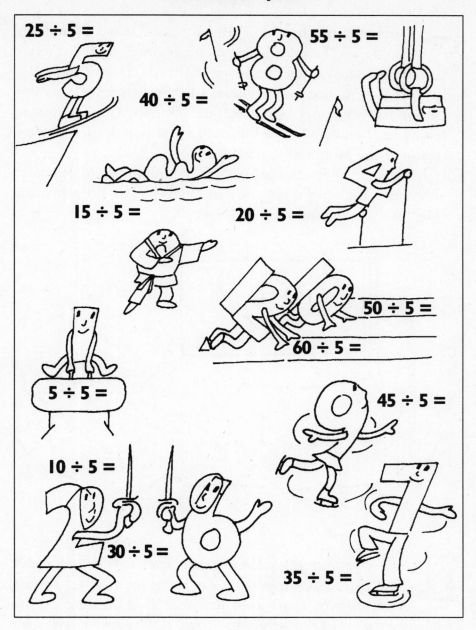

25 ÷ 5 =

55 ÷ 5 =

40 ÷ 5 =

15 ÷ 5 =

20 ÷ 5 =

50 ÷ 5 =

60 ÷ 5 =

5 ÷ 5 =

45 ÷ 5 =

10 ÷ 5 =

30 ÷ 5 =

35 ÷ 5 =

Notes for Myself . . . Just for Fun

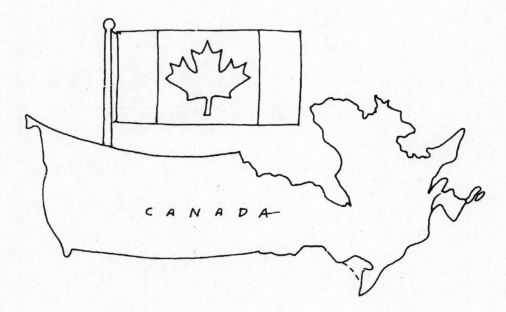

Canada is north of the United States. It has 26 million people and is split up into ten provinces and two territories. The capital of Canada is Ottawa. One of Canada's provinces, Quebec, was colonized by French settlers in the seventeenth century. Montreal and Quebec City are major cities in Quebec. One of Canada's most popular ice hockey teams is the Montreal Canadiens.*

While English is spoken throughout most of Canada, French is the official language in Quebec. The French word for "five" is *cinq* (pronounced SAYNK).

The next time you plan a family trip or vacation, enlist your child's participation. Let's say there are 5 people in the family. If 10 fruits are packed in an ice chest for the car trip, and everyone eats 1 fruit a day, how many days will fruits be available? Or, if it takes 6 hours to drive 300

*"Canada," *The Random House Children's Encyclopedia* (New York: Random House, 1991), pp. 89–91.

miles, how many miles are covered per hour? (If necessary, help your child see that 300 is a lot like 30.) Or, if there is $25 available to buy souvenirs, and everyone spends an equal amount, how much would each family member get to spend?

 My favorite number to divide by 5 is:

I like dividing by 5 because:

6. Flexible Six

Reptile/Dinosaur/Amphibian

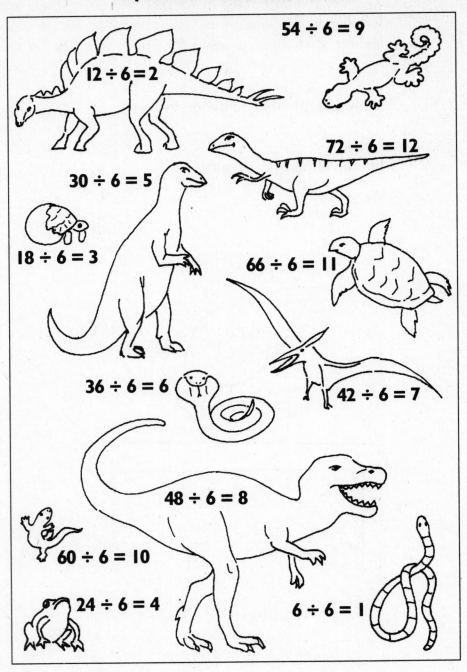

$54 \div 6 = 9$

$12 \div 6 = 2$

$72 \div 6 = 12$

$30 \div 6 = 5$

$18 \div 6 = 3$

$66 \div 6 = 11$

$36 \div 6 = 6$

$42 \div 6 = 7$

$48 \div 6 = 8$

$60 \div 6 = 10$

$24 \div 6 = 4$

$6 \div 6 = 1$

Individual Sports

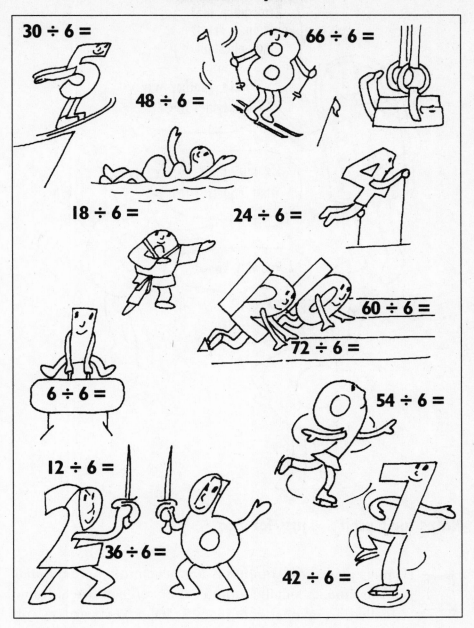

$30 \div 6 =$

$66 \div 6 =$

$48 \div 6 =$

$18 \div 6 =$

$24 \div 6 =$

$60 \div 6 =$

$72 \div 6 =$

$6 \div 6 =$

$54 \div 6 =$

$12 \div 6 =$

$36 \div 6 =$

$42 \div 6 =$

Notes for Myself . . . Just for Fun

Games played on computers are becoming more popular all the time. A small computer that you can use at home is called a personal computer. It has a keyboard so you can tell the computer what to do. It also has a screen, like a television screen, so you can see the game you're playing. Some games are played by typing at a keyboard; others by moving a joystick control or a mouse device.

There are all kinds of computer games you can play, and more are being invented every day. They let you pretend you're in places like a distant galaxy or an English castle or a martial arts tournament. If you could make up your own computer game, what would it be?

"The Input-Output Game." It's time to fire up your mental computer again. Can you write the correct output for each exercise below?

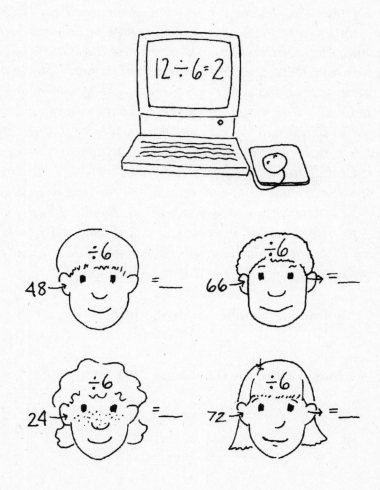

Six is a perfect number. What is a perfect number? It's a number that is the sum of all its divisors except the number itself. For 6, the divisors include 2, 3, and 1 (6 ÷ 2 = 3, 6 ÷ 1 = 6). And 6 is the sum of its divisors: 1 + 2 + 3 = 6. That's why 6 is a perfect number!

Tell your child that the next perfect number after 6 is 28. That is because the divisors of 28, which are 1, 2, 4, 7, and 14, all add up to 28.

 (1) When was the last time you cleaned out your closets? Get everyone in the family to collect all their old toys, clothes that no longer fit, and other items no longer in use, and donate them to a service center or other charitable organization. Have your child count up all the items of clothing (or toys, books, etc.), then ask him or her, "If the organization gives 6 items to every person, how many people will be helped?" For example, if your family donates 54 items, the child will divide by 6 to get the answer of 9, and tell you that 9 people would benefit. (2) If you donate money to a world relief fund, you could ask your child, "If it costs $6 to provide 1 child with immunizations for one year, and we give $24, how many children can we help?" If you donate time to a local relief agency you could ask, "If it takes 18 hours to paint a building, and we paint for 6 hours every week, how many weeks will it take to finish?"

My favorite number to divide by 6 is:

I like dividing by 6 because:

7. Seven on Stilts

Become a Mathlete! Exercise with Cal!

Reptile/Dinosaur/Amphibian

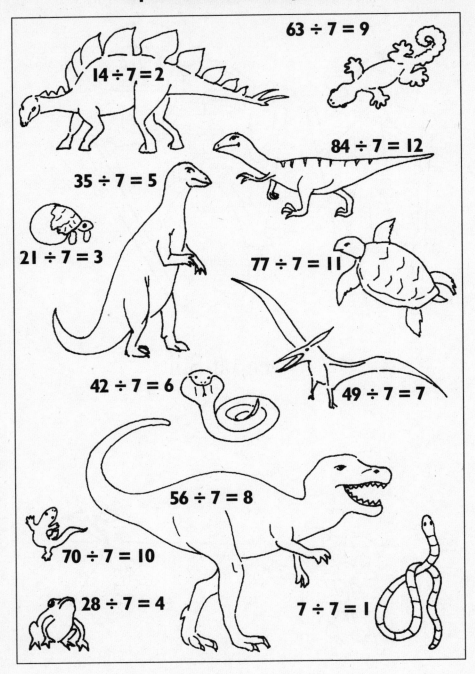

$63 \div 7 = 9$

$14 \div 7 = 2$

$84 \div 7 = 12$

$35 \div 7 = 5$

$21 \div 7 = 3$

$77 \div 7 = 11$

$42 \div 7 = 6$

$49 \div 7 = 7$

$56 \div 7 = 8$

$70 \div 7 = 10$

$28 \div 7 = 4$

$7 \div 7 = 1$

Individual Sports

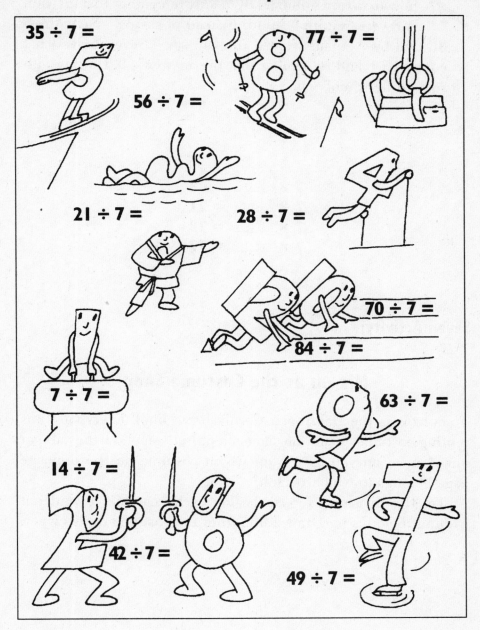

$35 \div 7 =$

$77 \div 7 =$

$56 \div 7 =$

$21 \div 7 =$

$28 \div 7 =$

$70 \div 7 =$

$84 \div 7 =$

$7 \div 7 =$

$63 \div 7 =$

$14 \div 7 =$

$42 \div 7 =$

$49 \div 7 =$

If you know how to multiply by 7, then you already know how to divide numbers by 7. Just remember that division is reverse multiplication. Instead of saying, "Sixty-three divided by 7 equals what?" you can say, "Seven times what equals 63?" and remember that the answer is 9. Do this with some other numbers:

$$7 \times \underline{\quad} = 49$$
$$7 \times \underline{\quad} = 21$$
$$7 \times \underline{\quad} = 28$$
$$7 \times \underline{\quad} = 77$$

Seven Activity:

Mix-up at the Costume Shop

Tomorrow is the Halloween Carnival at school. Everybody was renting costumes from the same shop, but somehow they all got mixed up. Can you figure out which costume on the opposite page belongs to which person?

Do the division exercise beside each costume to find out whom it belongs to. Draw a line from the costume to its owner.

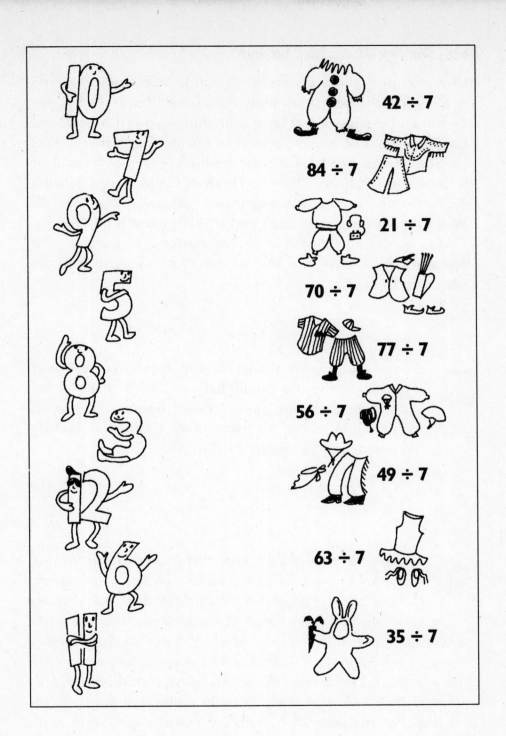

42 ÷ 7

84 ÷ 7

21 ÷ 7

70 ÷ 7

77 ÷ 7

56 ÷ 7

49 ÷ 7

63 ÷ 7

35 ÷ 7

Notes for Myself . . . Just for Fun

Halloween gives you a chance to pretend you're someone else. Try this game the next time you're with some friends: Everybody is to think of costumes that have something to do with numbers. Examples would be a mouse costume for the Three Blind Mice, or a Sinbad-type costume for the Seven Voyages of Sinbad or Ali Baba and the Forty Thieves. Think of books, songs, singing groups, poems, stories, rhymes, games, movies, plays, and TV shows for your ideas, but don't reveal them to each other.

After you've all thought of several costume ideas, give each other clues as to what they are. See who can come up with the most correct guesses.

Do you know where the word "Halloween" comes from? It means "the evening of all hallows," with "hallow" being Old English for "saint." That's because Halloween falls on the night before All Saints' Day, a Christian holiday that happens every November 1.

(1) Suppose your child empties the trash or feeds the dog every day in order to earn an allowance of $1 per week. If he or she has done this for 28 days, how much money has been earned? How about 42 days? If the child wants to save up $10, how many days would that be? (2) Assign someone in the family as "Birthday Monitor." This person's job is to circle all the special dates (birthdays, anniversaries, holidays) on a calendar, then to count down and keep you on track. For instance, if the next birthday in the family is 35 days away, the child divides by 7 to see how many weeks

away it is, and lets you know. At the end of the Birthday Monitor's service (e.g., three months), reward him or her in a special way, then rotate the assignment.

 My favorite number to divide by 7 is:

I like dividing by 7 because:

8. Skatin' Eight

Reptile/Dinosaur/Amphibian

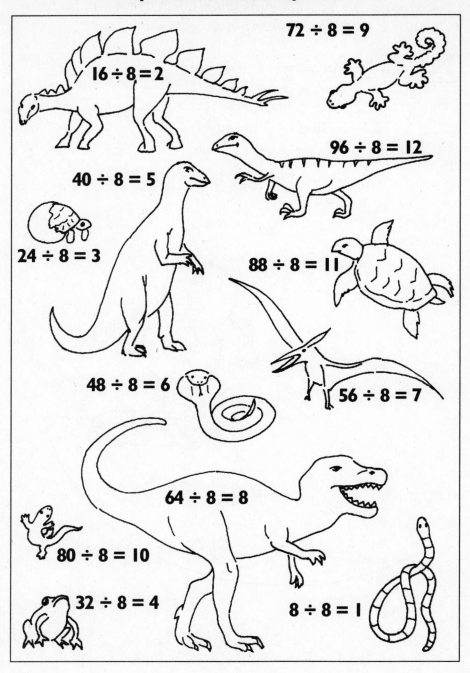

$72 \div 8 = 9$

$16 \div 8 = 2$

$96 \div 8 = 12$

$40 \div 8 = 5$

$24 \div 8 = 3$

$88 \div 8 = 11$

$48 \div 8 = 6$

$56 \div 8 = 7$

$64 \div 8 = 8$

$80 \div 8 = 10$

$32 \div 8 = 4$

$8 \div 8 = 1$

Individual Sports

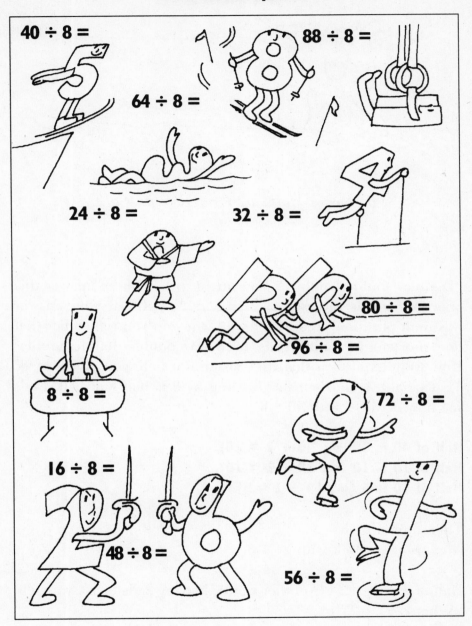

40 ÷ 8 =

88 ÷ 8 =

64 ÷ 8 =

24 ÷ 8 =

32 ÷ 8 =

80 ÷ 8 =

96 ÷ 8 =

8 ÷ 8 =

72 ÷ 8 =

16 ÷ 8 =

48 ÷ 8 =

56 ÷ 8 =

Become a Mathlete! Exercise with Cal!

"Double-Double-Double." You ought to be able to answer this one: Division is reverse _____ , right? If your answer was multiplication, hooray for you! Now, do you remember that one shortcut for multiplying by 8 was double-double-double? How would you turn that trick around to help you divide by 8? That's right, a shortcut for dividing by 8 is half-half-half. Let's see how it works with $40 \div 8$.

Half of 40 is 20 (or, $40 \div 2 = 20$).
Half of 20 is 10 (or, $20 \div 2 = 10$).
Half of 10 is 5 (or, $10 \div 2 = 5$).

Therefore, $40 \div 8 = 5$.
Now try it yourself with $24 \div 8$.

Half of 24 is _____ . (Another way of halving 24 is to ask yourself, "What is $24 \div 2$?")
Half of _____ is _____ .
Half of _____ is _____ .

$24 \div 8 =$ _____ .

Notes for Myself . . . Just for Fun

 Did you know that Hawaii is part of the United States? It became our fiftieth state in 1959. Hawaii is located in the Pacific Ocean and is a group of islands that actually are the tops of volcanoes. Some of these volcanoes are still active. Pineapples, sugarcane, and coffee beans are some of the main crops grown in Hawaii. It's also very popular with tourists.

How many original states were there in the United States? What's our forty-ninth state?

Get in the habit of visiting the library regularly with your child. If he or she can read, ask this question: "Say you have a book that is 32 pages long, and you read 8 pages per day. How many days will it take to finish the book?" Then pose the problem with other page counts that are divisible by 8. At that rate, which books would require more than a week to finish?

 My favorite number to divide by 8 is:

I like dividing by 8 because:

9. Doing Fine, Nine!

Division and multiplication go hand in hand. If you have done a division exercise and aren't sure if the answer is correct, just reverse it! Turn it into a multiplication exercise to check whether it's correct. For example, if you want to check whether $18 \div 9 = 2$ is correct, just reverse it by turning it into a multiplication exercise:

18 = 9 × 2

You know that's correct, so you know that your division equation is also correct.

Let's try a few others: $27 \div 9 = 3$, $45 \div 9 = 5$, and $72 \div 9 = 7$

27 = 9 × 3	CORRECT
45 = 9 × 5	CORRECT
72 = 9 × 7	OOPS!

So what's the correct answer for 72 ÷ 9? Ask yourself, "Nine multiplied by what equals 72?" to remember that the answer is 8.

72 = 9 × 8
9 × 8 = 72
72 ÷ 9 = 8

What's a quick way you can check whether a number can be divided by 9? Just see if the sum of its digits is 9, and if it is, then that number can be divided by 9. For example, you know that 54 can be divided by 9 because the sum of its digits, 5 and 4, add up to 9. Can 55 be divided by 9? Add its digits, 5 and 5, to get 10. So you cannot divide 55 by 9 (or at least, you can't divide it without having to break it apart into a fraction).

Circle the numbers below that can be divided by 9.

18
72
28
 9
34
81
45
36
40
63

You may be wondering about 99, one of the numbers that you're learning to divide by 9. At first, when you add up its digits, they don't seem to equal 9, because 9 + 9 = 18. But add these digits together one more time until you get a one-digit number, and you'll see that it works: 1 + 8 = 9.

Remember, this book uses whole numbers only.

Reptile/Dinosaur/Amphibian

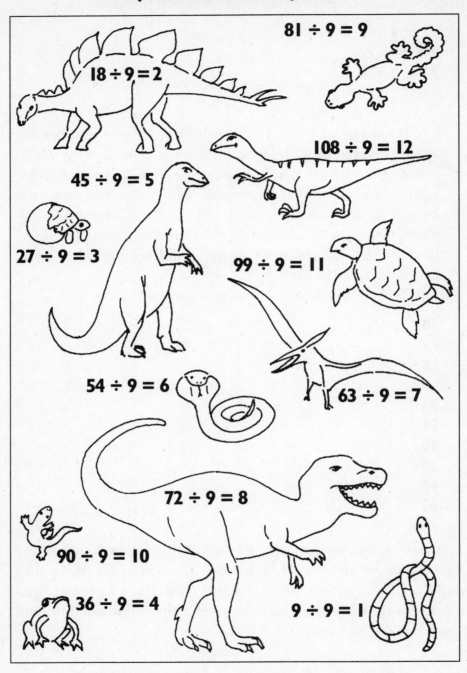

$81 \div 9 = 9$

$18 \div 9 = 2$

$108 \div 9 = 12$

$45 \div 9 = 5$

$27 \div 9 = 3$

$99 \div 9 = 11$

$54 \div 9 = 6$

$63 \div 9 = 7$

$72 \div 9 = 8$

$90 \div 9 = 10$

$36 \div 9 = 4$

$9 \div 9 = 1$

Division Squares

Remember Multiplication Squares from the last chapter? Division Squares are pretty much the same. Fill in any empty squares in the puzzle so that all the numbers fit into equations that work. Each equation will either read forward, or backward, or up, or down—but in only one direction. (None will work diagonally.) As you are filling in the missing numbers, draw arrows to indicate in what direction each equation should read.

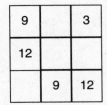

Some logic is required to solve these squares; help your children with them if necessary. Refer to the examples shown for the Multiplication Squares, in the discussion of 6 in the Multiplication chapter.

Reverse the division equations to come up with multiplication equations, and write these out to the side.

9		3
12		
	9	12

18		3
	2	
9		3

9		27
	3	
		3

Individual Sports

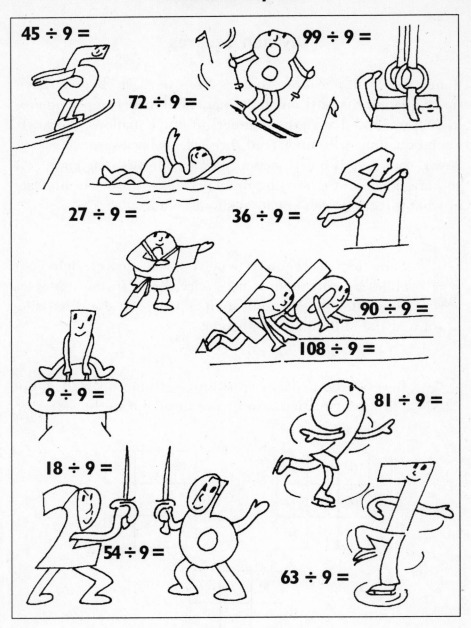

$45 \div 9 =$

$99 \div 9 =$

$72 \div 9 =$

$27 \div 9 =$

$36 \div 9 =$

$90 \div 9 =$

$108 \div 9 =$

$9 \div 9 =$

$81 \div 9 =$

$18 \div 9 =$

$54 \div 9 =$

$63 \div 9 =$

Notes for Myself . . . Just for Fun

The numbers we use today (0, 1, 2, 3 . . .) are called Arabic numbers. It is believed they were first used by the Hindu civilization about fourteen hundred years ago. An even older system of Roman numbers was used before then. The Romans had no number for 0, but here are some of the numbers they did have:*

I = 1	V = 5	IX = 9
II = 2	VI = 6	X = 10
III = 3	VII = 7	XI = 11
IV = 4	VIII = 8	XII = 12

Do you see any patterns in the Roman numerals above? Talk with your teacher or parents about what it means when I is to the left of V or X rather than to the right. Roman numerals are hardly ever used today except sometimes in book or movie titles. But they're fun to use when you're writing secret codes to your friends.

*"Mathematics," *The Random House Children's Encyclopedia* (New York: Random House, 1991), p. 335.

There are a number of worthwhile youth organizations that are rewarding for children in first grade and up, such as Boy Scouts, Girl Scouts, Campfire Girls, YMCA programs, and parks/recreation department programs. These offer ways your child can meet and share common interests with other children who may not attend the same school or live in the same neighborhood. Together with your child, plan a party for fellow group members. For example, ask your child, "If there are 9 members in your Cub Scout den, and you give a Mexican fiesta where you serve 27 tacos, how many tacos will each boy get?" Or, "If it takes you 9 days to earn a badge, and you work on different badge requirements for 45 days, how many badges will you earn?"

My favorite number to divide by 9 is:

I like dividing by 9 because:

Become a Mathlete! Exercise with Callie!

10. Hang Ten Again

Dividing by 10 is simple. For the numbers you will learn, the quotient is the same as the number you started out with—except with the 0 removed. For example:

$$10 \div 10 = 1$$
$$30 \div 10 = 3$$
$$50 \div 10 = 5$$
$$80 \div 10 = 8$$
$$100 \div 10 = 10$$

See how easy it is?

Ten Activity:

Don't Blow Your Top!

Callie and Ten are in Hawaii. They'd love to get to the coconut grove on the other side of this crater field. But some of these craters are unsafe because they're actually the tops of volcanoes that blow hot lava every now and then. Can you help them find a safe route?

Each of the volcano craters has a division equation on it. Some are right; some are wrong. Circle the ones that are right in order to get safely across.

Become a Mathlete! Exercise with Callie!

$70 \div 10 = 8$

$110 \div 10 = 11$

$10 \div 10 = 2$

$60 \div 10 = 6$

$10 \div 10 = 2$

$90 \div 10 = 9$

$50 \div 10 = 5$

$120 \div 10 = 2$

$100 \div 10 = 10$

$80 \div 10 = 8$

$120 \div 10 = 12$

$10 \div 10 = 1$

$40 \div 10 = 4$

$70 \div 10 = 7$

$20 \div 10 = 2$

$60 \div 10 = 5$

$30 \div 10 = 3$

$110 \div 10 = 10$

$100 \div 10 = 11$

Reptile/Dinosaur/Amphibian

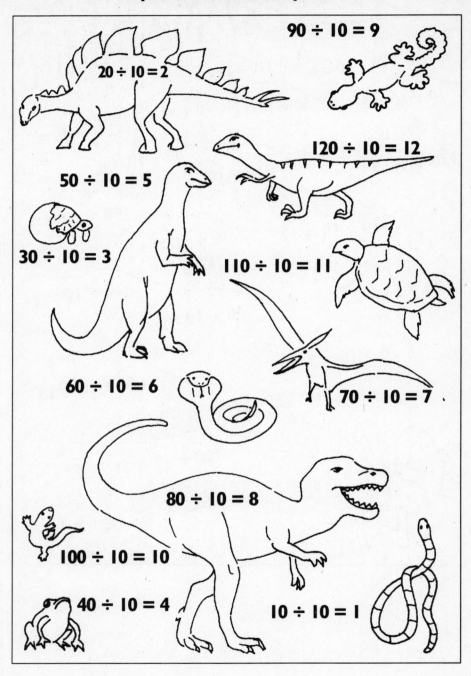

$90 \div 10 = 9$

$20 \div 10 = 2$

$120 \div 10 = 12$

$50 \div 10 = 5$

$110 \div 10 = 11$

$30 \div 10 = 3$

$60 \div 10 = 6$

$70 \div 10 = 7$

$80 \div 10 = 8$

$100 \div 10 = 10$

$40 \div 10 = 4$

$10 \div 10 = 1$

Individual Sports

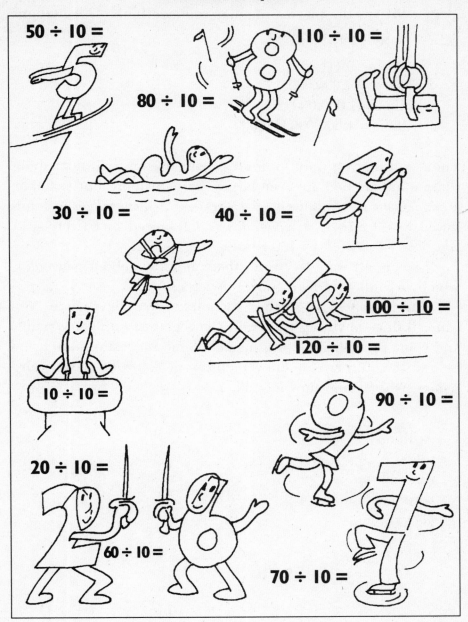

50 ÷ 10 =

110 ÷ 10 =

80 ÷ 10 =

30 ÷ 10 =

40 ÷ 10 =

100 ÷ 10 =

120 ÷ 10 =

10 ÷ 10 =

90 ÷ 10 =

20 ÷ 10 =

60 ÷ 10 =

70 ÷ 10 =

Notes for Myself . . . Just for Fun

 What's the biggest number you can think of? A million? A billion? A trillion?

1 million = 1,000,000
1 billion = 1,000,000,000
1 trillion = 1,000,000,000,000

The next time you want to impress your friends, tell them about some numbers that are even larger than a trillion. For example, a *quadrillion* is 1 followed by 15 zeroes. A *duodecillion* is 1 followed by 39 zeroes. A *vigintillion* is 1 followed by 63 zeroes. A *googol* is 1 followed by 100 zeroes.

If you really want to "wow" them, tell them about *googolplex*, which is 1 followed by a googol of zeroes!*

Since numbers bigger than googolplex don't have names, you can call them anything you want. Get together with your friends and come up with some wild names; some suggestions are listed below. If you were to draw pictures of some of these weird names, what would they look like?

gazillion
jillion
dinomillion
sprocketillion
geekazillion

*Theoni Pappas, "Watch Out for the Googols," in *Fractals, Googols and Other Mathematical Tales* (San Carlos, Calif.: Wide World Publishing/Tetra, 1993), pp. 16–17.

The last time you were on a car trip with your family, did you ever wonder how long it would take to count to a really big number, like 100 or 1,000? Here's one way to do it, now that you know multiplication and division: First, decide how long it takes to say each number—3 seconds is a good guess. Then multiply 3 seconds by the number you want to count to, such as 100.

100 × 3 = 300

Now you know that it will take 300 seconds to count to 100. To find out how long that it is in minutes, use division. Since there are 60 seconds in a minute, you would divide by 60.

300 ÷ 60 = 5

And there you have it—it takes 5 minutes (or 300 seconds) to count to 100!

(Remember, even if you don't feel sure about dividing by 60, you *do* know how to divide by 6. If you cross out one of the 0s from 300, and one of the 0s from 60, then you have 30 ÷ 6, and you certainly know the answer to that one! In other words, the quotient for 300 ÷ 60 is the same as for 30 ÷ 6. The answer is 5.)

Now that you know it takes 5 minutes to count to 100, how long do you suppose it would take to count to 1,000? Well, you could go through the same steps as you went through before. Or, you could use an even faster way. First, know that 1,000 is 10 times bigger than 100. So what number is 10 times bigger than 5? That's right; the answer is 50. It would take 50 minutes to count to 1,000.

Look at this table, and fill in the missing numbers:

100 ↔ 5 minutes
1,000 ↔ 5__ minutes
10,000 ↔ 50__ minutes
100,000 ↔ 5,00__ minutes
1,000,000 ↔ 50,00__ minutes

 For bigger numbers, such as 500 or 5,000 minutes, help your child learn how to convert to hours (60 minutes in an hour) or days (24 hours in a day).

(1) Help your child become comfortable with dividing large numbers by 10. Write a series of such numbers on paper (e.g., 1,000; 10,000; 3,000; 453,000), and cross out the last 0 to show what the number becomes after it has been divided by 10. Do this many times until your child is confident. (2) Reinforce your child in learning the names of very large numbers. Together, come up with everyday examples of things that occur in very large numbers, such as the number of stars in the sky, or grains of sand on a beach, or insects in the world. Then ask such questions as "If there are 1,000,000 grains of sand on a beach, how many times would you have to divide them by 10 before you had only 100 grains?" or "If you start out with 5,000 bugs, how many times would you divide by 10 to end up with only a handful of bugs?" Have your child write the original number down, then count the number of times he or she crosses out a 0 to arrive at the desired number.

 My favorite number to divide by 10 is:

I like dividing by 10 because:

11. Fun-Loving Eleven

Become a Mathlete! Exercise with Callie!

Reptile/Dinosaur/Amphibian

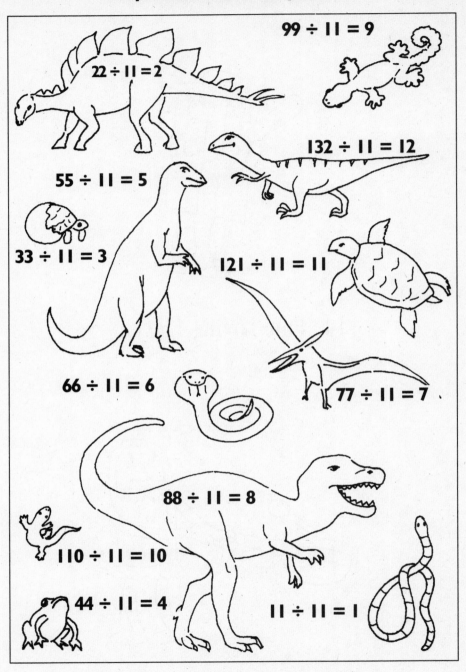

$99 \div 11 = 9$

$22 \div 11 = 2$

$132 \div 11 = 12$

$55 \div 11 = 5$

$33 \div 11 = 3$

$121 \div 11 = 11$

$66 \div 11 = 6$

$77 \div 11 = 7$

$88 \div 11 = 8$

$110 \div 11 = 10$

$44 \div 11 = 4$

$11 \div 11 = 1$

Eleven Activity:

Straight from the Horse's Mouth

Callie is working in the Beanbag Toss booth at the school carnival. A list tells her what prize to award, depending on how many beanbags a player tosses through the horse's mouth. But someone has replaced most of the numbers on the list with division problems. Can you help her solve them?

First, solve any of the following that are division equations:

Number of Beanbags Accurately Thrown	Prize Awarded
$77 \div 11$	bead necklace
3	whistle
$11 \div 11$	balloon
$55 \div 11$	ring
$22 \div 11$	top
8	pencil
$132 \div 11$	stuffed animal
$99 \div 11$	yo-yo
$44 \div 11$	fake bug
10	jigsaw puzzle
$66 \div 11$	fake nose
$121 \div 11$	checkers game

Now, rearrange and write in the numbers and prizes below, starting with 1 beanbag and its corresponding prize.

Number of Beanbags Accurately Thrown	Prize Awarded
1	

Individual Sports

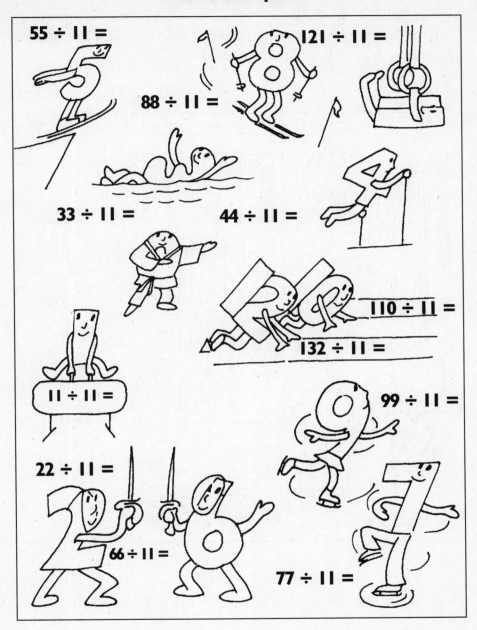

55 ÷ 11 =

121 ÷ 11 =

88 ÷ 11 =

33 ÷ 11 =

44 ÷ 11 =

110 ÷ 11 =

132 ÷ 11 =

11 ÷ 11 =

99 ÷ 11 =

22 ÷ 11 =

66 ÷ 11 =

77 ÷ 11 =

Notes for Myself . . . Just for Fun

Next to Callie is the Balloon Pop booth, where Eleven is working. He just compared his prize chest with Callie's and found that her prize chest has 11 items that his does not.

Find the 11 missing objects, which are hidden somewhere in this picture.

Encourage your child to help the family in planning and sticking to a budget. For example, if it will cost $99 to buy a new bookcase, and the family has 11 weeks in which to put aside money for the purchase, how much should be saved each week? If it will cost $132 to buy new clothes for school, with the same time period of 11 weeks in which to save money, how much should be put aside each week for that purpose? What if your child wants to buy a CD-ROM that costs $33? Discuss with your child how the saving period can be shortened if a greater amount of money is put aside each week, and how that would affect the rest of the family budget that is available for other expenses. If your child is old enough, encourage him or her to contribute to the family's earning power by doing chores for neighbors, such as mowing the lawn or baby-sitting.

My favorite number to divide by 11 is:

I like dividing by 11 because:

12. Twelve on the Trapeze

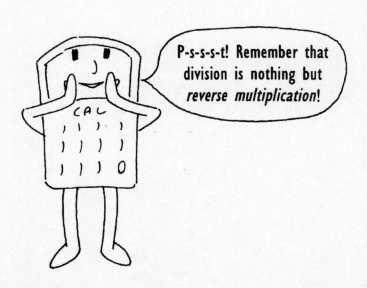

Reptile/Dinosaur/Amphibian

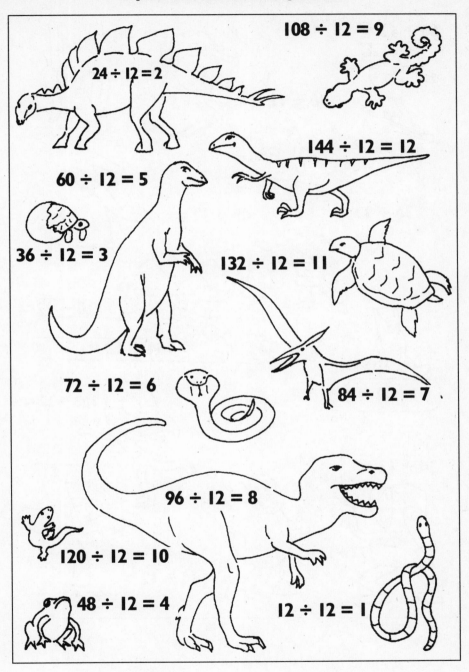

$108 \div 12 = 9$

$24 \div 12 = 2$

$144 \div 12 = 12$

$60 \div 12 = 5$

$36 \div 12 = 3$

$132 \div 12 = 11$

$72 \div 12 = 6$

$84 \div 12 = 7$

$96 \div 12 = 8$

$120 \div 12 = 10$

$48 \div 12 = 4$

$12 \div 12 = 1$

Individual Sports

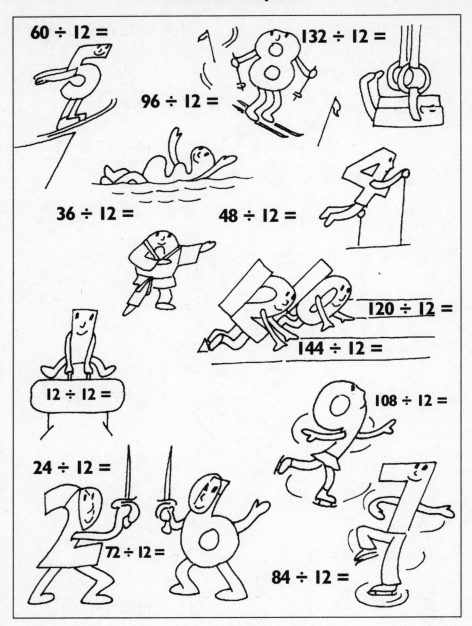

60 ÷ 12 =

132 ÷ 12 =

96 ÷ 12 =

36 ÷ 12 =

48 ÷ 12 =

120 ÷ 12 =

144 ÷ 12 =

12 ÷ 12 =

108 ÷ 12 =

24 ÷ 12 =

72 ÷ 12 =

84 ÷ 12 =

Notes for Myself . . . Just for Fun

Birthdays are important days in our lives. Did you know that people live longer nowadays than they used to?

There are lots of reasons why; one is that we've learned more about what makes us healthy and are taking better care of ourselves.

The *life span* of a plant or animal is how long it typically lives. For example, the life span of a human being right now is about seventy-five years—which is certainly longer than that of a butterfly, but shorter than certain other plants and animals. In fact, you'd probably be surprised if you knew what some common life spans are.*

Look at the life spans below in Column A, then match them up with the plants or animals in Column B. Turn page 322 upside down to compare your guesses with the actual answers. How close were you?

Column A	Column B
20 minutes	goldfish
11,000 years	morning-glory flower
90 years	bedbug
8 years	killer whale
6 months	queen ant
15 years	some bacteria
3 weeks	red fox
1 day	giant sequoia
120 years	creosote bush
6,000 years	tortoise
700 years	housefly
30 years	ponderosa pine

*"Plant and Animal Lifespans," *The Random House Children's Encyclopedia* (New York: Random House, 1991), p. 616.

(1) If your child has a chart on which to post stickers earned for doing chores, that's a good opportunity to reinforce division by 12. For example, say the chart is sectioned into 12 parts, 1 for each month. Ask your child, "If there are 72 stickers on this chart so far, and an equal number of stickers for each month, how many stickers did you earn each month?" If every 12 stickers add up to a special treat, ask, "If you've collected 60 stickers, how many prizes have you earned?" (2) Recycle household items to use in storing supplies. For example, an empty egg carton is ideal for collecting small sewing items such as buttons, or office supplies such as paper clips, or art supplies such as thumb tacks. Say your child wants to use the egg carton for collecting stickers. If he or she wants to redistribute a box of 84 stickers into the egg carton, how many would go into each compartment? Or if you give your child 96 beads, with an equal number to go in each compartment, have him or her figure out beforehand what the correct number is.

My favorite number to divide by 12 is:

I like dividing by 12 because:

Answers

bacteria (20 minutes)
morning-glory flower (1 day)
housefly (3 weeks)
bedbug (6 months)
red fox (8 years)
queen ant (15 years)
goldfish (30 years)
killer whale (90 years)
tortoise (120 years)
ponderosa pine (700 years)
giant sequoia (6,000 years)
creosote bush (11,000 years)

Division Review

Now you're an expert at division! Isn't it fun? It's kind of like the "flip side" of multiplication. In fact, that's what division is: *reverse multiplication*. When you ask yourself something like, "Ten divided by 5 equals what?" that's the same as asking yourself, "Five multiplied by what equals 10?" In either case, the answer is 2.

In division, the answer is called a *quotient*. What's great about it is that it lets you split things up evenly. For instance, you know that if you and your friend have 6 pieces of gum to share, each of you gets 3 pieces, because $6 \div 2 = 3$.

Even though 0 is an important number in multiplication, it is not used in division. However, just as in multiplication, when you *divide* any number by 1, it stays the same. (This also means that if you divide any number by itself, the quotient is 1.)

When you divide a number by 2, you get *half* the original amount. When you divide any number by 4, just think of it as "half-half." And when you divide any number by 8—that's right—it's "half-half-half."

Finally, dividing any number by 10 is almost as easy as dividing it by 1. The quotient will always be the same as the number you started out with, except with the 0 removed.

Become a Mathlete! Exercise with Cal!

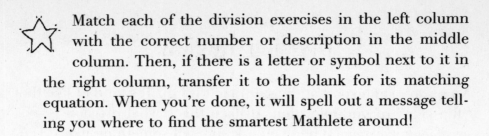

Match each of the division exercises in the left column with the correct number or description in the middle column. Then, if there is a letter or symbol next to it in the right column, transfer it to the blank for its matching equation. When you're done, it will spell out a message telling you where to find the smartest Mathlete around!

40 ÷ 5 ___	planets in our solar system	I
11 ÷ 11 ___	11	R
84 ÷ 12	quarters in a dollar	M
33 ÷ 11 ___	arms on a starfish	R
144 ÷ 12	days in a week	
36 ÷ 9 ___	ears on your head	R
54 ÷ 6 ___	1	N
132 ÷ 12 ___	eggs in a carton	
16 ÷ 8 ___	notes in an octave	I
48 ÷ 8 ___	dimes in a dollar	!
60 ÷ 12 ___	sides of a triangle	A
100 ÷ 10 ___	6	O

Answer Key

Chapter 3: Putting It All Together

Page 17
Color the Zerosaurus Rex: $0 + 1 = 1$, $0 + 2 = 2$, $0 + 3 = 3$, $0 + 4 = 4$, $0 + 5 = 5$, $0 + 6 = 6$, $0 + 7 = 7$, $0 + 8 = 8$

Pages 21–22
Count Your Toys: $1 + 5 = 6$, $8 + 1 = 9$, $6 + 1 = 7$, $1 + 3 = 4$, $4 + 1 = 5$

Page 23
Schoolyard Games: $0 + 1 = 1$, $1 + 1 = 2$, $2 + 1 = 3$, $3 + 1 = 4$, $4 + 1 = 5$, $5 + 1 = 6$, $6 + 1 = 7$, $7 + 1 = 8$, $8 + 1 = 9$, $9 + 1 = 10$, $10 + 1 = 11$, $11 + 1 = 12$, $12 + 1 = 13$

Page 26
Two by Two: 14 animals

Page 27
Schoolyard Games: $0 + 2 = 2$, $1 + 2 = 3$, $2 + 2 = 4$, $3 + 2 = 5$, $4 + 2 = 6$, $5 + 2 = 7$, $6 + 2 = 8$, $7 + 2 = 9$, $8 + 2 = 10$, $9 + 2 = 11$, $10 + 2 = 12$, $11 + 2 = 13$, $12 + 2 = 14$

Page 31
Schoolyard Games: $0 + 3 = 3$, $1 + 3 = 4$, $2 + 3 = 5$, $3 + 3 = 6$, $4 + 3 = 7$, $5 + 3 = 8$, $6 + 3 = 9$, $7 + 3 = 10$, $8 + 3 = 11$, $9 + 3 = 12$, $10 + 3 = 13$, $11 + 3 = 14$, $12 + 3 = 15$

Page 34
"Grouping": 6, 8

Page 37
Hit the Bases: second, third, home

Page 39
Just for Fun: Using only two numbers at a time, there are three ways to add up to 5 ($0 + 5$ or $5 + 0$, $1 + 4$ or $4 + 1$, and $2 + 3$ or $3 + 2$); four ways to add up to 6 ($0 + 6$ or $6 + 0$, $1 + 5$ or $5 + 1$, $2 + 4$ or $4 + 2$, and $3 + 3$); four ways to add up to 7 ($0 + 7$ or $7 + 0$, $1 + 6$ or $6 + 1$, $2 + 5$ or $5 + 2$, and $3 + 4$ or $4 + 3$); five ways to add up to 8 ($0 + 8$ or $8 + 0$, $1 + 7$ or $7 + 1$, $2 + 6$ or $6 + 2$, $3 + 5$ or $5 + 3$, and $4 + 4$); and five ways to add up to 9 ($0 + 9$ or $9 + 0$, $1 + 8$ or $8 + 1$, $2 + 7$ or $7 + 2$, $3 + 6$ or $6 + 3$, and $4 + 5$ or $5 + 4$). Here is the pattern: 4 (3 ways), 5 (3 ways), 6 (4 ways), 7 (4 ways), 8 (5 ways), 9 (5 ways), and so on. Following this pattern, 10 would have six ways.

Page 45

Schoolyard Games: 0 + 6 = 6, 1 + 6 = 7, 2 + 6 = 8, 3 + 6 = 9,
4 + 6 = 10, 5 + 6 = 11, 6 + 6 = 12, 7 + 6 = 13, 8 + 6 = 14,
9 + 6 = 15, 10 + 6 = 16, 11 + 6 = 17, 12 + 6 = 18

Page 49

More "Grouping": 14

Page 52

A Few of My Favorite Things: 12, 9, 10

Page 53

Schoolyard Games: 0 + 7 = 7, 1 + 7 = 8, 2 + 7 = 9, 3 + 7 = 10,
4 + 7 = 11, 5 + 7 = 12, 6 + 7 = 13, 7 + 7 = 14, 8 + 7 = 15,
9 + 7 = 16, 10 + 7 = 17, 11 + 7 = 18, 12 + 7 = 19

Page 58

Don't Be Late — Race with Eight!: Cars labeled 0 + 8, 7 + 8, 8 + 2,
8 + 5, 9 + 8 should be circled

Page 59

Schoolyard Games: 0 + 8 = 8, 1 + 8 = 9, 2 + 8 = 10, 3 + 8 = 11,
4 + 8 = 12, 5 + 8 = 13, 6 + 8 = 14, 7 + 8 = 15, 8 + 8 = 16,
9 + 8 = 17, 10 + 8 = 18, 11 + 8 = 19, 12 + 8 = 20

Page 63

Crazy Craters: 9 + 3 = 12, 9 + 0 = 9, 9 + 1 = 10, 4 + 9 = 13,
12 + 9 = 21, and 6 + 9 = 15

Page 64

Schoolyard Games: 0 + 9 = 9, 1 + 9 = 10, 2 + 9 = 11, 3 + 9 = 12,
4 + 9 = 13, 5 + 9 = 14, 6 + 9 = 15, 7 + 9 = 16, 8 + 9 = 17,
9 + 9 = 18, 10 + 9 = 19, 11 + 9 = 20, 12 + 9 = 21

Page 66

"Nine Squares": There is more than one solution. Here is one:

2	4	3
4	3	2
3	2	4

Page 68

More "Grouping": 27

Pages 70–71

Wax On, Wax Off: 11, 16, 19, 15, 22. The equation for 19 should be
circled.

Page 72

Schoolyard Games: 0 + 10 = 10, 1 + 10 = 11, 2 + 10 = 12,
3 + 10 = 13, 4 + 10 = 14, 5 + 10 = 15, 6 + 10 = 16, 7 + 10 = 17,

$8 + 10 = 18, 9 + 10 = 19, 10 + 10 = 20, 11 + 10 = 21,$
$12 + 10 = 22$

Page 74
Zeros: 50, 90, 100

Page 77
Martian Message:

$$\underset{12}{\text{M}} \underset{14}{\text{A}} \underset{5}{\text{T}} \underset{17}{\text{H}} \qquad \underset{20}{\text{I}} \underset{13}{\text{S}} \qquad \underset{23}{\text{O}} \underset{15}{\text{U}} \underset{5}{\text{T}} \qquad \underset{23}{\text{O}} \underset{11}{\text{F}}$$

$$\underset{5}{\text{T}} \underset{17}{\text{H}} \underset{20}{\text{I}} \underset{13}{\text{S}} \qquad \underset{6}{\text{W}} \underset{23}{\text{O}} \underset{8}{\text{R}} \underset{18}{\text{L}} \underset{4}{\text{D}} \,!$$

Page 78
Schoolyard Games: $0 + 11 = 11, 1 + 11 = 12, 2 + 11 = 13,$
$3 + 11 = 14, 4 + 11 = 15, 5 + 11 = 16, 6 + 11 = 17, 7 + 11 = 18,$
$8 + 11 = 19, 9 + 11 = 20, 10 + 11 = 21, 11 + 11 = 22,$
$12 + 11 = 23$

Page 82
Schoolyard Games: $0 + 12 = 12, 1 + 12 = 13, 2 + 12 = 14,$
$3 + 12 = 15, 4 + 12 = 16, 5 + 12 = 17, 6 + 12 = 18, 7 + 12 = 19,$
$8 + 12 = 20, 9 + 12 = 21, 10 + 12 = 22, 11 + 12 = 23,$
$12 + 12 = 24$

Page 86
Left-to-Right Addition: 87

Page 91
Extra Credit:

1. $7 + 1 < 6 + 3 \ (8 < 9)$
2. $12 + 4 > 7 + 8 \ (16 > 15)$
3. $10 + \text{wheels on a bicycle} = 6 + 6 \ (12 = 12)$
4. $4 + 9 = 6 + 7 \ (13 = 13)$
5. $2 + 8 < 7 + 4 \ (10 < 11)$
6. $1 + 5 > 3 + \text{eyes on your face} \ (6 > 5)$
7. $0 + 11 < 9 + 3 \ (11 < 12)$
8. $10 + 9 > 6 + 12 \ (19 > 18)$
9. $2 + 1 = 1 + 2 \ (3 = 3)$
10. $9 + 7 > 11 + 3 \ (16 > 14)$
11. $\text{legs on a hippo} + 4 > 2 + 5 \ (8 > 7)$
12. $5 + 8 < 7 + 7 \ (13 < 14)$
13. $8 + 10 > 6 + 11 \ (18 > 17)$
14. $4 + 2 = 5 + 1 \ (6 = 6)$
15. $2 + \text{doughnuts in a dozen} < 4 + 11 \ (14 < 15)$
16. $10 + 6 < 9 + 8 \ (16 < 17)$
17. $0 + 3 > 1 + 1 \ (3 > 2)$
18. $6 + 4 > 8 + 1 \ (10 > 9)$

19. $9 + 11 = 13 + 7$ $(20 = 20)$
20. $5 + 9 >$ months in a year $+ 1$ $(14 > 13)$

Pages 91–92
Addition Table Activities:

1.   12
2.   0
3.   The answer depends on your age.
4.   Start with the 0 in the upper-left-hand corner, then progress diagonally to the 2 in the second row, then to the 4 in the third row, and so on. All of these are even numbers.
5.   The sum of the 2s column is 13 more than the sum of the 1s column. The sum of the 3s column will also be 13 more than the sum of the 2s column, and so on for each successive column.

Chapter 4: Take It Away, Cal!

Page 96
Closet Cleanup: 2, 3, 5, 1, 4, 4,792, *xyz*

Page 100
Team Sports: $1 - 1 = 0, 2 - 1 = 1, 3 - 1 = 2, 4 - 1 = 3,$
$5 - 1 = 4, 6 - 1 = 5, 7 - 1 = 6, 8 - 1 = 7, 9 - 1 = 8, 10 - 1 = 9,$
$11 - 1 = 10, 12 - 1 = 11, 13 - 1 = 12$

Page 104
Time Two Leave? $2 - 2 = 0, 3 - 2 = 1, 4 - 2 = 2, 5 - 2 = 3,$
$6 - 2 = 4$

Page 105
Team Sports: $2 - 2 = 0, 3 - 2 = 1, 4 - 2 = 2, 5 - 2 = 3,$
$6 - 2 = 4, 7 - 2 = 5, 8 - 2 = 6, 9 - 2 = 7, 10 - 2 = 8,$
$11 - 2 = 9, 12 - 2 = 10, 13 - 2 = 11, 14 - 2 = 12$

Page 109
Backward Bowling: 7, 5, 4, 1, 6, 3, Cal

Page 110
Team Sports: $3 - 3 = 0, 4 - 3 = 1, 5 - 3 = 2, 6 - 3 = 3,$
$7 - 3 = 4, 8 - 3 = 5, 9 - 3 = 6, 10 - 3 = 7, 11 - 3 = 8,$
$12 - 3 = 9, 13 - 3 = 10, 14 - 3 = 11, 15 - 3 = 12$

Page 111
Just for Fun: blind mice, little pigs, bears

Page 115
Team Sports: $4 - 4 = 0, 5 - 4 = 1, 6 - 4 = 2, 7 - 4 = 3,$
$8 - 4 = 4, 9 - 4 = 5, 10 - 4 = 6, 11 - 4 = 7, 12 - 4 = 8,$
$13 - 4 = 9, 14 - 4 = 10, 15 - 4 = 11, 16 - 4 = 12$

Page 119

Up and Down: sixth, fifth, two, two, third floor, fifth floor

Page 120

Team Sports: $5 - 5 = 0$, $6 - 5 = 1$, $7 - 5 = 2$, $8 - 5 = 3$, $9 - 5 = 4$, $10 - 5 = 5$, $11 - 5 = 6$, $12 - 5 = 7$, $13 - 5 = 8$, $14 - 5 = 9$, $15 - 5 = 10$, $16 - 5 = 11$, $17 - 5 = 12$

Page 124

Crack the Code:

$$\underset{3}{\text{I}} \quad \underset{12}{\text{W}} \, \underset{1}{\text{A}} \, \underset{6}{\text{N}} \, \underset{2}{\text{T}} \quad \underset{4}{\text{M}} \, \underset{10}{\text{Y}} \quad \underset{4}{\text{M}} \, \underset{9}{\text{U}} \, \underset{4}{\text{M}} \, \underset{4}{\text{M}} \, \underset{10}{\text{Y}}$$

Page 125

Team Sports: $6 - 6 = 0$, $7 - 6 = 1$, $8 - 6 = 2$, $9 - 6 = 3$, $10 - 6 = 4$, $11 - 6 = 5$, $12 - 6 = 6$, $13 - 6 = 7$, $14 - 6 = 8$, $15 - 6 = 9$, $16 - 6 = 10$, $17 - 6 = 11$, $18 - 6 = 12$

Page 129

Team Sports: $7 - 7 = 0$, $8 - 7 = 1$, $9 - 7 = 2$, $10 - 7 = 3$, $11 - 7 = 4$, $12 - 7 = 5$, $13 - 7 = 6$, $14 - 7 = 7$, $15 - 7 = 8$, $16 - 7 = 9$, $17 - 7 = 10$, $18 - 7 = 11$, $19 - 7 = 12$

Page 130

Extra Credit: All these subtraction exercises have 7 as their answer.

Page 133

Team Sports: $8 - 8 = 0$, $9 - 8 = 1$, $10 - 8 = 2$, $11 - 8 = 3$, $12 - 8 = 4$, $13 - 8 = 5$, $14 - 8 = 6$, $15 - 8 = 7$, $16 - 8 = 8$, $17 - 8 = 9$, $18 - 8 = 10$, $19 - 8 = 11$, $20 - 8 = 12$

Page 138

Team Sports: $9 - 9 = 0$, $10 - 9 = 1$, $11 - 9 = 2$, $12 - 9 = 3$, $13 - 9 = 4$, $14 - 9 = 5$, $15 - 9 = 6$, $16 - 9 = 7$, $17 - 9 = 8$, $18 - 9 = 9$, $19 - 9 = 10$, $20 - 9 = 11$, $21 - 9 = 12$

Page 141

Ring Around the Collar: $10 - 1$, 3, $10 - 8$, 11, $10 + 1$, 10

Page 143

Team Sports: $10 - 10 = 0$, $11 - 10 = 1$, $12 - 10 = 2$, $13 - 10 = 3$, $14 - 10 = 4$, $15 - 10 = 5$, $16 - 10 = 6$, $17 - 10 = 7$, $18 - 10 = 8$, $19 - 10 = 9$, $20 - 10 = 10$, $21 - 10 = 11$, $22 - 10 = 12$

Page 144

Magic Act: 10, 20, 40, 30, 50

Page 148

Crazy Classroom: Only two equations on the chalkboard are correct: $15 - 11 = 4$, and $22 - 11 = 11$. The correct answers to the others should be: $18 - 11 = 7$, $12 - 11 = 1$, $21 - 11 = 10$, $16 - 11 = 5$, and $11 - 11 = 0$.

Page 149
Team Sports: 11 − 11 = 0, 12 − 11 = 1, 13 − 11 = 2, 14 − 11 = 3,
15 − 11 = 4, 16 − 11 = 5, 17 − 11 = 6, 18 − 11 = 7, 19 − 11 = 8,
20 − 11 = 9, 21 − 11 = 10, 22 − 11 = 11, 23 − 11 = 12

Page 150
Just for Fun: 1. ice hockey; 2. bowling; 3. football; 4. tennis

Page 153
Landlubbers Ahoy:

12 − 12 =	0	N
16 − 12 =	4	V
20 − 12 =	8	T
23 − 12 =	11	F
16 − 13 =	3	P
13 − 12 =	1	H
12 − 12 =	0	U
12 − 9 =	3	R
7 − 5 =	2	E
19 − 12 =	7	S

$$\underline{H}\ \underline{A}\ \underline{V}\ \underline{E} \qquad \underline{F}\ \underline{U}\ \underline{N}$$
$$\underline{1}\ \underline{2}\ \underline{4}\ \underline{5} \qquad \underline{11}\ \underline{0}\ \underline{12}$$

Page 154
Team Sports: 12 − 12 = 0, 13 − 12 = 1, 14 − 12 = 2, 15 − 12 = 3,
16 − 12 = 4, 17 − 12 = 5, 18 − 12 = 6, 19 − 12 = 7, 20 − 12 = 8,
21 − 12 = 9, 22 − 12 = 10, 23 − 12 = 11, 24 − 12 = 12

Chapter 5: Growing by Leaps and Bounds

Page 163
The Winning Way: 4 × 0 = 0, 5 × 0 = 0, 6 × 0 = 0, 7 × 0 = 0,
8 × 0 = 0, 9 × 0 = 0, 10 × 0 = 0, 11 × 0 = 0, 12 × 0 = 0

Page 164
Camp Activity: 0 × 0 = 0, 1 × 0 = 0, 2 × 0 = 0, 3 × 0 = 0,
4 × 0 = 0, 5 × 0 = 0, 6 × 0 = 0, 7 × 0 = 0, 8 × 0 = 0, 9 × 0 = 0,
10 × 0 = 0, 11 × 0 = 0, 12 × 0 = 0

Page 169
Go Fly a Kite!:

Multiplication Equation	True	False
1 × 1 = 2	A	E
1 × 8 = 8	R	C
4 × 1 = 1	Y	Z
1 × 1 = 1	B	D
10 × 1 = 10	E	P
2 × 1 = 3	G	E

Page 170
Camp Activity: $0 \times 1 = 0$, $1 \times 1 = 1$, $2 \times 1 = 2$, $3 \times 1 = 3$, $4 \times 1 = 4$, $5 \times 1 = 5$, $6 \times 1 = 6$, $7 \times 1 = 7$, $8 \times 1 = 8$, $9 \times 1 = 9$, $10 \times 1 = 10$, $11 \times 1 = 11$, $12 \times 1 = 12$

Pages 174–175
Up in the Rafters: b, a, b, a, c, a, c

Page 176
Camp Activity: $0 \times 2 = 0$, $1 \times 2 = 2$, $2 \times 2 = 4$, $3 \times 2 = 6$, $4 \times 2 = 8$, $5 \times 2 = 10$, $6 \times 2 = 12$, $7 \times 2 = 14$, $8 \times 2 = 16$, $9 \times 2 = 18$, $10 \times 2 = 20$, $11 \times 2 = 22$, $12 \times 2 = 24$

Page 177
Extra Credit: C

Page 181
Extra Credit:

3	6	**9**	12	**15**	18	**21**	**24**	27
30	**33**	36	**39**	42	**45**	**48**	51	**54**
57	**60**	**63**	66	**69**	**72**	75	**78**	81
84	**87**	90	**93**	**96**	99	**102**	**105**	108

Page 182
Camp Activity: $0 \times 3 = 0$, $1 \times 3 = 3$, $2 \times 3 = 6$, $3 \times 3 = 9$, $4 \times 3 = 12$, $5 \times 3 = 15$, $6 \times 3 = 18$, $7 \times 3 = 21$, $8 \times 3 = 24$, $9 \times 3 = 27$, $10 \times 3 = 30$, $11 \times 3 = 33$, $12 \times 3 = 36$

Page 187
Four Corners Cuisine:

4×1	5	S		4×7	28	I
4×2	8	C		4×8	30	N
4×3	14	P		4×9	36	L
4×4	**16**	H		4×10	40	I
4×5	25	D		4×11	42	M
4×6	26	F		4×12	50	Y

C H I L I

Page 188
Fun Fact: Arizona, Colorado, New Mexico, Utah

Page 189
Camp Activity: $0 \times 4 = 0$, $1 \times 4 = 4$, $2 \times 4 = 8$, $3 \times 4 = 12$, $4 \times 4 = 16$, $5 \times 4 = 20$, $6 \times 4 = 24$, $7 \times 4 = 28$, $8 \times 4 = 32$, $9 \times 4 = 36$, $10 \times 4 = 40$, $11 \times 4 = 44$, $12 \times 4 = 48$

Page 195

Camp Activity: $0 \times 5 = 0$, $1 \times 5 = 5$, $2 \times 5 = 10$, $3 \times 5 = 15$, $4 \times 5 = 20$, $5 \times 5 = 25$, $6 \times 5 = 30$, $7 \times 5 = 35$, $8 \times 5 = 40$, $9 \times 5 = 45$, $10 \times 5 = 50$, $11 \times 5 = 55$, $12 \times 5 = 60$

Pages 200–201

A Dicey Situation:

1. 30
2. 12
3. 24
4. 18
5. 42
6. 54
7. 72

Page 203

Camp Activity: $0 \times 6 = 0$, $1 \times 6 = 6$, $2 \times 6 = 12$, $3 \times 6 = 18$, $4 \times 6 = 24$, $5 \times 6 = 30$, $6 \times 6 = 36$, $7 \times 6 = 42$, $8 \times 6 = 48$, $9 \times 6 = 54$, $10 \times 6 = 60$, $11 \times 6 = 66$, $12 \times 6 = 72$

Page 206

Multiplication Squares:

2	3	6
6	2	3
12	6	2

1	4	4
2	3	6
2	12	24

Page 210

Christmas in the Big Apple: $7 \times 2 = 14$

Page 210

Extra Credit: $7 \times 3 = 21$, $25 - 21 = 4$, therefore today is December 4.

Page 211

Camp Activity: $0 \times 7 = 0$, $1 \times 7 = 7$, $2 \times 7 = 14$, $3 \times 7 = 21$, $4 \times 7 = 28$, $5 \times 7 = 35$, $6 \times 7 = 42$, $7 \times 7 = 49$, $8 \times 7 = 56$, $9 \times 7 = 63$, $10 \times 7 = 70$, $11 \times 7 = 77$, $12 \times 7 = 84$

Page 212

Grouping: 9, 12, $9 + 12 = 21$; 3, 18, $3 + 18 = 21$

Page 218

Camp Activity: $0 \times 8 = 0$, $1 \times 8 = 8$, $2 \times 8 = 16$, $3 \times 8 = 24$, $4 \times 8 = 32$, $5 \times 8 = 40$, $6 \times 8 = 48$, $7 \times 8 = 56$, $8 \times 8 = 64$, $9 \times 8 = 72$, $10 \times 8 = 80$, $11 \times 8 = 88$, $12 \times 8 = 96$

Page 220

Stargazing:

$9 \times 5 =$	**45 (P)**	40 (L)	50 (O)
$9 \times 4 =$	39 (S)	**36 (R)**	34 (T)
$9 \times 8 =$	81 (H)	76 (G)	**72 (I)**
$9 \times 3 =$	18 (R)	**27 (T)**	28 (W)
$9 \times 9 =$	90 (L)	**81 (J)**	82 (I)
$9 \times 6 =$	53 (C)	56 (D)	**54 (E)**
$9 \times 7 =$	**63 (U)**	64 (Y)	66 (V)

JUPITER

Page 223

Camp Activity: $0 \times 9 = 0$, $1 \times 9 = 9$, $2 \times 9 = 18$, $3 \times 9 = 27$, $4 \times 9 = 36$, $5 \times 9 = 45$, $6 \times 9 = 54$, $7 \times 9 = 63$, $8 \times 9 = 72$, $9 \times 9 = 81$, $10 \times 9 = 90$, $11 \times 9 = 99$, $12 \times 9 = 108$

Page 225

Extra Credit: Mercury, Venus, Earth, Mars, Jupiter, Saturn, Uranus, Neptune, Pluto; 1. Jupiter; 2. Mercury; 3. Mars; 4. Jupiter, Saturn, Uranus, Neptune; 5. Mars and Jupiter; 6. Earth, Mars, Jupiter, Saturn, Neptune, Pluto

Page 229

Camp Activity: $0 \times 10 = 0$, $1 \times 10 = 10$, $2 \times 10 = 20$, $3 \times 10 = 30$, $4 \times 10 = 40$, $5 \times 10 = 50$, $6 \times 10 = 60$, $7 \times 10 = 70$, $8 \times 10 = 80$, $9 \times 10 = 90$, $10 \times 10 = 100$, $11 \times 10 = 110$, $12 \times 10 = 120$

Page 234

Camp Activity: $0 \times 11 = 0$, $1 \times 11 = 11$, $2 \times 11 = 22$, $3 \times 11 = 33$, $4 \times 11 = 44$, $5 \times 11 = 55$, $6 \times 11 = 66$, $7 \times 11 = 77$, $8 \times 11 = 88$, $9 \times 11 = 99$, $10 \times 11 = 110$, $11 \times 11 = 121$, $12 \times 11 = 132$

Page 235

Magic Act: <u>1 2</u>, $1 + 2 = 3$, <u>132</u>

Page 240

Camp Activity: $0 \times 12 = 0$, $1 \times 12 = 12$, $2 \times 12 = 24$, $3 \times 12 = 36$, $4 \times 12 = 48$, $5 \times 12 = 60$, $6 \times 12 = 72$, $7 \times 12 = 84$, $8 \times 12 = 96$, $9 \times 12 = 108$, $10 \times 12 = 120$, $11 \times 12 = 132$, $12 \times 12 = 144$

Page 241

Extra Credit: 12, 1, 24, 2, 60, 60, 5

Page 242

Grouping: 40, 8, $40 + 8 = 48$

Page 246
Extra Credit:

1. $4 \times 10 = 8 \times 5$ (40 = 40)
2. $5 \times 3 > 2 \times 7$ (15 > 14)
3. $9 \times 6 < 7 \times 8$ (54 < 56)
4. $3 \times 11 > 5 \times$ sides on a cube (33 > 30)
5. $6 \times 8 < 7 \times 7$ (48 < 49)
6. $2 \times 9 = 3 \times 6$ (18 = 18)
7. $10 \times 10 < 12 \times 12$ (100 < 144)
8. $3 \times 3 > 2 \times 4$ (9 > 8)
9. eggs in a dozen $< 7 \times 2$ (12 < 14)
10. $9 \times 3 > 5 \times 5$ (27 > 25)
11. $12 \times 9 < 11 \times 10$ (108 < 110)
12. $5 \times 12 = 10 \times 6$ (60 = 60)
13. legs on a horse $\times 11 > 6 \times$ days in a week (44 > 42)
14. $6 \times 6 = 3 \times 12$ (36 = 36)
15. $1 \times 1 > 0 \times 2$ (1 > 0)
16. arms on an octopus $\times 8 > 9 \times 7$ (64 > 63)
17. $3 \times 9 < 4 \times 7$ (27 < 28)
18. $11 \times 11 > 12 \times 10$ (121 > 120)
19. $8 \times 3 = 6 \times 4$ (24 = 24)
20. $12 \times 6 = 9 \times 8$ (72 = 72)

Chapter 6: Share and Share Alike

Page 252
Individual Sports: $1 \div 1 = 1, 2 \div 1 = 2, 3 \div 1 = 3, 4 \div 1 = 4,$
$5 \div 1 = 5, 6 \div 1 = 6, 7 \div 1 = 7, 8 \div 1 = 8, 9 \div 1 = 9,$
$10 \div 1 = 10, 11 \div 1 = 11, 12 \div 1 = 12$

Page 253
Extra Credit: 1, 1, 1

Page 258
Let's Halve a Ton of Fun:

1.	1	7.	12
2.	3	8.	6
3.	11	9.	7
4.	9	10.	4
5.	5	11.	8
6.	2	12.	10

Page 259
Individual Sports: $2 \div 2 = 1, 4 \div 2 = 2, 6 \div 2 = 3, 8 \div 2 = 4,$

$10 \div 2 = 5$, $12 \div 2 = 6$, $14 \div 2 = 7$, $16 \div 2 = 8$, $18 \div 2 = 9$,
$20 \div 2 = 10$, $22 \div 2 = 11$, $24 \div 2 = 12$

Page 264
Individual Sports: $3 \div 3 = 1$, $6 \div 3 = 2$, $9 \div 3 = 3$, $12 \div 3 = 4$,
$15 \div 3 = 5$, $18 \div 3 = 6$, $21 \div 3 = 7$, $24 \div 3 = 8$, $27 \div 3 = 9$,
$30 \div 3 = 10$, $33 \div 3 = 11$, $36 \div 3 = 12$

Page 266
Extra Credit: 10, 12, 7

Page 268
Magic Act: 10, 10, 5, 5

Page 270
Guessing Game:

1. 11
2. 6
3. 9
4. 8

Page 271
Individual Sports: $4 \div 4 = 1$, $8 \div 4 = 2$, $12 \div 4 = 3$, $16 \div 4 = 4$,
$20 \div 4 = 5$, $24 \div 4 = 6$, $28 \div 4 = 7$, $32 \div 4 = 8$, $36 \div 4 = 9$,
$40 \div 4 = 10$, $44 \div 4 = 11$, $48 \div 4 = 12$

Page 272
Fun Fact: $100 \div 4 = 25$

Page 276
Neighbor to the North: $50 \div 5 = 10$, $15 \div 5 = 3$, $60 \div 5 = 12$,
$25 \div 5 = 5$, $40 \div 5 = 8$

Page 277
Individual Sports: $5 \div 5 = 1$, $10 \div 5 = 2$, $15 \div 5 = 3$, $20 \div 5 = 4$,
$25 \div 5 = 5$, $30 \div 5 = 6$, $35 \div 5 = 7$, $40 \div 5 = 8$, $45 \div 5 = 9$,
$50 \div 5 = 10$, $55 \div 5 = 11$, $60 \div 5 = 12$

Page 281
Individual Sports: $6 \div 6 = 1$, $12 \div 6 = 2$, $18 \div 6 = 3$, $24 \div 6 = 4$,
$30 \div 6 = 5$, $36 \div 6 = 6$, $42 \div 6 = 7$, $48 \div 6 = 8$, $54 \div 6 = 9$,
$60 \div 6 = 10$, $66 \div 6 = 11$, $72 \div 6 = 12$

Page 283
"The Input-Output Game": 8, 4, 12, 11

Page 287
Individual Sports: $7 \div 7 = 1$, $14 \div 7 = 2$, $21 \div 7 = 3$, $28 \div 7 = 4$,
$35 \div 7 = 5$, $42 \div 7 = 6$, $49 \div 7 = 7$, $56 \div 7 = 8$, $63 \div 7 = 9$,
$70 \div 7 = 10$, $77 \div 7 = 11$, $84 \div 7 = 12$

Page 288
Magic Act: 7, 3, 4, 11

Page 289

Mix-up at the Costume Shop:

$42 \div 7 \rightarrow$ (Six Character)
$21 \div 7 \rightarrow$ (Three Character)
$49 \div 7 \rightarrow$ (Seven Character)
$35 \div 7 \rightarrow$ (Five Character)
$84 \div 7 \rightarrow$ (Twelve Character)
$70 \div 7 \rightarrow$ (Ten Character)
$56 \div 7 \rightarrow$ (Eight Character)
$63 \div 7 \rightarrow$ (Nine Character)
$77 \div 7 \rightarrow$ (Eleven Character)

Page 293

Individual Sports: $8 \div 8 = 1$, $16 \div 8 = 2$, $24 \div 8 = 3$, $32 \div 8 = 4$, $40 \div 8 = 5$, $48 \div 8 = 6$, $56 \div 8 = 7$, $64 \div 8 = 8$, $72 \div 8 = 9$, $80 \div 8 = 10$, $88 \div 8 = 11$, $96 \div 8 = 12$

Page 294

"Double-Double-Double": 12, 12, 6, 6, 3, 3

Page 295

Extra Credit: Alaska

Page 297

Magic Act: The following numbers from the list should be circled: 18, 72, 9, 81, 45, 36, 63.

Page 299

Division Squares:

9	3	3
12	3	4
108	9	12

18	6	3
2	2	1
9	3	3

9	3	27
3	3	9
3	1	3

Page 299

Extra Credit for Division Squares:
$3 \times 3 = 9$, $9 \times 12 = 108$, $3 \times 4 = 12$, $12 \times 9 = 108$, $4 \times 3 = 12$
$3 \times 6 = 18$, $2 \times 1 = 2$, $3 \times 3 = 9$, $9 \times 2 = 18$, $2 \times 3 = 6$, $3 \times 1 = 3$
$3 \times 3 = 9$, $3 \times 1 = 3$, $1 \times 3 = 3$, $3 \times 9 = 27$, $9 \times 3 = 27$

Page 300

Individual Sports: $9 \div 9 = 1$, $18 \div 9 = 2$, $27 \div 9 = 3$, $36 \div 9 = 4$, $45 \div 9 = 5$, $54 \div 9 = 6$, $63 \div 9 = 7$, $72 \div 9 = 8$, $81 \div 9 = 9$, $90 \div 9 = 10$, $99 \div 9 = 11$, $108 \div 9 = 12$

Page 305

Don't Blow Your Top!: $10 \div 10 = 1$, $20 \div 10 = 2$, $30 \div 10 = 3$,

$40 \div 10 = 4$, $50 \div 10 = 5$, $60 \div 10 = 6$, $70 \div 10 = 7$, $80 \div 10 = 8$, $90 \div 10 = 9$, $100 \div 10 = 10$, $110 \div 10 = 11$, $120 \div 10 = 12$

Page 307
Individual Sports: $10 \div 10 = 1$, $20 \div 10 = 2$, $30 \div 10 = 3$, $40 \div 10 = 4$, $50 \div 10 = 5$, $60 \div 10 = 6$, $70 \div 10 = 7$, $80 \div 10 = 8$, $90 \div 10 = 9$, $100 \div 10 = 10$, $110 \div 10 = 11$, $120 \div 10 = 12$

Page 310
Toughie:

$1,000 = 50$ minutes

$10,000 = 500$ minutes (or 8.33 hours)

$100,000 = 5,000$ minutes (or 83.33 hours) (or 3.47 days)

$1,000,000 = 50,000$ minutes (or 833 hours) (or 34.7 days) (or over a month!)

Pages 313–314
Straight from the Horse's Mouth:

Number of Beanbags Accurately Thrown	Prize Awarded
$77 \div 11$ **7**	bead necklace
3	whistle
$11 \div 11$ **1**	balloon
$55 \div 11$ **5**	ring
$22 \div 11$ **2**	top
8	pencil
$132 \div 11$ **12**	stuffed animal
$99 \div 11$ **9**	yo-yo
$44 \div 11$ **4**	fake bug
10	jigsaw puzzle
$66 \div 11$ **6**	fake nose
$121 \div 11$ **11**	checkers game

Number of Beanbags Accurately Thrown	Prize Awarded
1	balloon
2	top
3	whistle
4	fake bug
5	ring
6	fake nose
7	bead necklace
8	pencil
9	yo-yo
10	jigsaw puzzle
11	checkers game
12	stuffed animal

Page 315
Individual Sports: $11 \div 11 = 1$, $22 \div 11 = 2$, $33 \div 11 = 3$, $44 \div 11 = 4$, $55 \div 11 = 5$, $66 \div 11 = 6$, $77 \div 11 = 7$, $88 \div 11 = 8$, $99 \div 11 = 9$, $110 \div 11 = 10$, $121 \div 11 = 11$, $132 \div 11 = 12$

Page 316
Balloon Pop Booth: bead necklace, whistle, balloon, ring, top, pencil, stuffed animal, yo-yo, fake bug, fake nose, checkers game

Page 320
Individual Sports: $12 \div 12 = 1$, $24 \div 12 = 2$, $36 \div 12 = 3$, $48 \div 12 = 4$, $60 \div 12 = 5$, $72 \div 12 = 6$, $84 \div 12 = 7$, $96 \div 12 = 8$, $108 \div 12 = 9$, $120 \div 12 = 10$, $132 \div 12 = 11$, $144 \div 12 = 12$

Page 324
Extra Credit:

$40 \div 5$ _I_	planets in our solar system	I
$11 \div 11$ _N_	11	R
$84 \div 12$	quarters in a dollar	M

33 ÷ 11 __A__	arms on a starfish	R
144 ÷ 12	days in a week	R
36 ÷ 9 __M__	ears on your head	R
54 ÷ 6 __I__	1	N
132 ÷ 12 __R__	eggs in a carton	
16 ÷ 8 __R__	notes in an octave	I
48 ÷ 8 __O__	dimes in a dollar	!
60 ÷ 12 __R__	sides of a triangle	A
100 ÷ 10 __!__	6	O